Air Force Readiness Assessment

How Training Infrastructure Can Provide Better
Information for Decisionmaking

EMMI YONEKURA, DAVID SCHULKER, IRINA A. CHINDEA,
AJAY K. KOCHHAR, ANDREA M. ABLER, MARK TOUKAN, MATTHEW WALSH

Prepared for the Department of the Air Force
Approved for public release; distribution unlimited

RAND | PROJECT AIR FORCE

For more information on this publication, visit **www.rand.org/t/RRA992-2**.

About RAND

The RAND Corporation is a research organization that develops solutions to public policy challenges to help make communities throughout the world safer and more secure, healthier and more prosperous. RAND is nonprofit, nonpartisan, and committed to the public interest. To learn more about RAND, visit www.rand.org.

Research Integrity

Our mission to help improve policy and decisionmaking through research and analysis is enabled through our core values of quality and objectivity and our unwavering commitment to the highest level of integrity and ethical behavior. To help ensure our research and analysis are rigorous, objective, and nonpartisan, we subject our research publications to a robust and exacting quality-assurance process; avoid both the appearance and reality of financial and other conflicts of interest through staff training, project screening, and a policy of mandatory disclosure; and pursue transparency in our research engagements through our commitment to the open publication of our research findings and recommendations, disclosure of the source of funding of published research, and policies to ensure intellectual independence. For more information, visit www.rand.org/about/research-integrity.

RAND's publications do not necessarily reflect the opinions of its research clients and sponsors.

Published by the RAND Corporation, Santa Monica, Calif.
© 2023 RAND Corporation
RAND® is a registered trademark.

Library of Congress Cataloging-in-Publication Data is available for this publication.
ISBN: 978-1-9774-1222-5

Cover: U.S. Air Force photo by Javier Garcia.

About This Report

The objective of the U.S. Air Force's training program is to deliver readiness by building and sustaining operator skills and to provide information for assessing the readiness of individuals and teams and, ultimately, of the joint force. Yet, senior U.S. Department of Defense leadership is increasingly concerned that the current readiness assessment system is not providing sufficient insight into the capability of the force to meet future mission requirements—that there is a shortfall in the quality of inputs and, therefore, the outputs of the readiness system. If the Air Force makes appropriate investments, its training infrastructure, which in total is referred to as the *operational test and training infrastructure* (OTTI), could provide much more insight into the readiness of the force for future contingencies. This report examines and characterizes shortfalls in the readiness assessment process and then reviews potential OTTI remedies. Discussions with senior leaders at several major commands, review of current design plans for a Common Synthetic Training Environment, and a literature review of technological developments informed this analysis.

The research reported here was commissioned by Headquarters Air Force A3T and conducted within the Workforce, Development, and Health Program of RAND Project AIR FORCE as part of a fiscal year 2021 project "Operational Training Infrastructure and Live, Virtual, and Constructive Environments in Support of Squadron Commander Assessments of Unit Readiness." A companion report from this same project defines different dimensions of OTTI and the current state of technologies across those dimensions.[1]

RAND Project AIR FORCE

RAND Project AIR FORCE (PAF), a division of the RAND Corporation, is the Department of the Air Force's (DAF's) federally funded research and development center for studies and analyses, supporting both the United States Air Force and the United States Space Force. PAF provides the DAF with independent analyses of policy alternatives affecting the development, employment, combat readiness, and support of current and future air, space, and cyber forces. Research is conducted in four programs: Strategy and Doctrine; Force Modernization and Employment; Resource Management; and Workforce, Development, and Health. The research reported here was prepared under contract FA7014-16-D-1000.

---

[1] Mark Toukan, Matthew Walsh, Ajay K. Kochhar, Emmi Yonekura, and David Schulker, *Air Force Operational Test and Training Infrastructure: Barriers to Full Implementation*, RAND Corporation, RR-A992-1, 2022.

Additional information about PAF is available on our website:
www.rand.org/paf/

This report documents work originally shared with the DAF on October 5, 2021. The draft report, dated September 2021, was reviewed by formal peer reviewers and DAF subject-matter experts.

Acknowledgments

We would like to thank many people who were involved with and supported the research presented in this report. We are grateful to our Air Force sponsor, Steven Ruehl, the Deputy Director of Training and Readiness at Headquarters Air Force, for his support throughout the project. We also thank Lillian Campbell-Wynn, the Live, Virtual, Constructive Operations Advisor at Air Force Agency for Modeling and Simulation, for her contributions, especially in coordinating with many offices across the Air Force.

We are very grateful to the Air Force training and readiness stakeholders whose views informed the research conducted for this project, especially those from Air Combat Command, Air Mobility Command, Air Force Global Strike Command, and Air Force Special Operations Command.

Additionally, John Butcher and John Diercks from AF/A3TR have our gratitude for sharing their insights into the readiness assessment process and for connecting us with different readiness stakeholders. We thank Stephen Mark Webb, Sonia R. Vonderlippe, and Allan J. Fluharty from Air Force Agency for Modeling and Simulation for connecting us with the latest developments on the Common Synthetic Training Environment.

We would like to thank our RAND colleagues, Timothy Marler, Barbara Bicksler, Anthony Rosello, and Anna Jean Wirth, who improved the report with constructive reviews. We thank Raymond Conley for project guidance and Tara Terry for helping identify live, virtual, constructive subject-matter experts. We are also grateful to Nelson Lim, director of RAND Project AIR FORCE's Workforce, Development and Health Program, and Kirsten Keller and Miriam Matthews, the associate directors of the Workforce, Development and Health Program, who supported us throughout the research project.

Summary

Issue

Senior Department of the Air Force leadership is increasingly concerned that the current readiness assessment system is not providing sufficient insight into the capability of the force to meet future mission requirements because of the lack of quality outcome measurements in the readiness system. Concurrently, the U.S. Air Force (USAF) is evolving its training infrastructure in response to the prospect of operations in contested and denied environments, an increased pace of warfare, and the potential loss of superiority across multiple domains in a conflict with near-peer adversaries. Advances in the technological capabilities of training infrastructure can help fill gaps in current readiness assessments to provide senior leaders with better insight into the readiness of the force for future contingencies.

Approach

To understand how investments in training infrastructure could fill gaps in readiness assessment, we used a multimethod qualitative approach that included an extensive review of the relevant bodies of literature; policy and USAF documents, including Defense Readiness Reporting System squadron reports; and training system technical documents. We also conducted 13 discussions with four Air Force senior leaders and nine technical experts or subject-matter experts on readiness from various major commands. Ultimately, the synthesis of our analysis yielded recommended options for the future design of training infrastructure that take into account the benefits to readiness assessment.

Key Findings

- Leaders across the Department of Defense need readiness assessments that consider the ability of disparate military units to integrate and conduct the full spectrum of operations against any adversary. The USAF senior leaders we interviewed think about readiness along two dimensions: resource readiness and capability readiness. Adjustments are needed for resource and capability readiness to align with the needs of Department of Defense leaders.
- The USAF is not measuring the most useful things to gain insight on the readiness of the force. Legacy metrics focus on the ability of individual service members to conduct individual missions. But most National Defense Strategy missions require an integrated approach: Both USAF training requirements and how training is achieved need to change to capture more-meaningful readiness metrics.
- We identified three gaps in the current readiness assessment process: (1) measurement of factors that come into play only when forces are integrated, (2) readiness report

aggregation that does not match force presentation, and (3) the requirement for unit commanders to report readiness on threat environments and scenarios they cannot or rarely train against. These gaps cannot be addressed using the current training infrastructure.

- Senior leaders we interviewed across major commands identified the following investments in training assets to help address readiness assessment gaps: (1) distributed mission operations training; (2) more simulators in general; (3) new synthetic threat environments; (4) aggregated force readiness measurement; and (5) adaptive, proficiency-driven training.
- The Air Force plans for the new Common Synthetic Training Environment contain technical challenges in the design process; the decisions made to resolve these challenges will affect how well the new system will improve readiness assessment.

Recommendations

- **Further differentiate capability readiness and align new dimensions with supporting inputs from appropriate functions at headquarters and major commands.** The Air Force should define specific, measurable, attainable, relevant, and time-bound (SMART) elements of the broader definition of *readiness* and align these elements with inputs that can be provided by appropriate functional organizations across the service (e.g., inputs from the Intelligence Directorate on adversary capabilities and from the Logistics Directorate for issues of sustaining capabilities in extended scenarios).
- **Consider a process mechanism to bring information into readiness reporting from more-appropriate sources when unit commanders lack information.** We recommend that planners use the best available information from across different functions to inform readiness reporting. Then, the same functional areas can leverage new synthetic training opportunities to improve the state of capability knowledge and, in turn, improve future readiness assessments.
- **Consider adding a field in Defense Readiness Reporting System–Strategic to capture the quality of information used as inputs for subjective assessments.** Adding such a field would be an immediate improvement to the data-collection approach. Collecting this information explicitly would provide feedback on the quality of information informing subjective assessments today. More importantly, however, it would position the Air Force to measure the impact of new synthetic training capabilities on the quality of information flowing into the system.
- **Create a working group focused on data and measurement to guide synthetic environment design decisions.** A wide range of entities stands to benefit from the general-purpose information that might be created by future synthetic training environments. To ensure that new synthetic environments meet the diverse needs of these stakeholders, the Air Force should form a semipermanent cross-functional working group to advise acquisition efforts on design issues pertaining to data and measurement.
- **Factor readiness assessment gaps into operational test and training infrastructure (OTTI) priorities.** Plans and priorities for future OTTI capabilities might not realize the full benefit of the capabilities unless they also factor in the impact of training technologies on readiness assessment gaps. Planning documents, such as the OTTI Flight

Plan, should consider the readiness benefits when setting priorities for OTTI development.

Contents

Figures and Tables

Figures

Tables

1. Introduction

Background

National security senior leaders face a difficult and complex planning environment. The National Defense Strategy (NDS) states that the "central challenge to U.S. prosperity and security is the *reemergence of long-term, strategic competition*," with the "revisionist powers" of China and Russia.[2] Discussions about this strategic environment often express alarm at the degree to which these rivals have narrowed the capability advantages that the United States has enjoyed for most of the post–Cold War era.[3]

While these challenges demand a significant response, congressional policymakers have also signaled a desire for senior national security leaders to make the many necessary changes to capability investments and organization by reprioritizing within current resource levels rather than dramatically increasing budgets for defense. For instance, the House Armed Services Committee *Future of Defense Task Force Report 2020* notes that federal budgets will probably shrink while these national security challenges grow, necessitating a new paradigm that takes a "broad view of what investments are considered to be critical to the nation's security, as well as hard choices about how to apportion increasingly limited resources."[4]

The requirement to reorganize around a potential future conflict with more-capable adversaries has brought the topic of military readiness to the forefront of policy discussions. The quality of our understanding of how readiness is generated is a key ingredient in the planning process because the only way to effectively prioritize investments is to understand how bundles of investments will contribute to capabilities and, ultimately, to prevailing in future conflicts. Yet, readiness metrics have a long history of inadequacy because of the difficulty of understanding how groups of individuals and pieces of equipment will perform in complex, never-before-observed, operational environments.[5] Furthermore, there is a kind of observer effect because the act of testing readiness (e.g., flying a training sortie) can actually consume readiness. The added difficulty in an environment of strategic competition is that planners cannot know when a conflict might break out or how long it might last. Despite this uncertainty, policies still must determine a portfolio of investments that mature according to varying time horizons: Investing too much in long-term capabilities places near-term objectives at risk and vice versa.

---

[2] Jim Mattis, *Summary of the 2018 National Defense Strategy of the United States of America: Sharpening the American Military's Competitive Edge*, 2018, p. 2; emphasis in the original.

[3] Mattis, 2018; Seth Moulton, Jim Banks, Susan Davis, Scott DesJarlais, Chrissy Houlahan, Paul Mitchell, Elissa Slotkin, and Michael Waltz, *Future of Defense Task Force Report 2020*, House Armed Services Committee, 2020.

[4] Moulton et al., 2020, p. 63.

[5] Richard K. Betts, *Military Readiness: Concepts, Choices, Consequences*, Brookings Institution, 1995.

Senior U.S. Department of Defense (DoD) and Department of the Air Force (DAF) leaders acknowledge the flaws and inherent difficulties of understanding and measuring readiness but, at the same time, express hope that new analytic methods can produce new decision-quality information for planning. In a recent article on the need to redefine readiness, the Chief of Staff of the Air Force Gen. Charles Q. Brown, Jr., suggested that, by "using big data, machine learning, and AI [artificial intelligence], we should be able to build a model that accurately reflects these [historically missing] elements."[6] Similarly, Secretary of Defense Lloyd Austin, in his written responses to policy questions prior to his nomination, said that he would improve readiness assessments by seeking "to employ advances in the fields of data science to make our data more strategically informative."[7] These comments raise a question: If the consensus is that historical readiness metrics are not informative, what data will inform new quantitative models of readiness?

In this report, we discuss the ongoing evolution of readiness definitions and metrics and explore one potential source of nontraditional data that could inform new quantitative models of readiness: synthetic training environments within the operational test and training infrastructure (OTTI). Because training often involves rehearsing capabilities in simulated combat, it is a natural source of information on readiness. Technology for data capture and the increasing feasibility of performing more-realistic simulated exercises within synthetic training environments present opportunities to glean new information on readiness without additional costs to personnel or equipment (beyond what is needed for training).

Research Approach

To understand how synthetic training environments could potentially fill gaps in readiness measurement, we first analyzed the existing readiness system and identified shortfalls.[8] Then, we explored ways in which future synthetic training environments might yield opportunities to mitigate these shortfalls in the context of other emerging technological enablers. Finally, we examined the synthetic environment design specifications that determine the extent to which planners can realize these opportunities. We relied on a multimethod qualitative approach to research that included (1) discussions with USAF senior leaders, technical experts, and subject-matter experts (SMEs) on readiness from various major commands (MAJCOMs) relevant to the project and (2) extensive review of the existing literature on readiness, aircrew training and

---

[6] Charles Q. Brown, Jr., and David H. Berger, "Redefine Readiness or Lose," War on the Rocks, March 15, 2021.

[7] Lloyd J. Austin, "Senate Armed Services Committee Advance Policy Questions for Lloyd J. Austin Nominee for Appointment to be Secretary of Defense," January 19, 2021, p. 12.

[8] As part of this research, the current state of OTTI in the U.S. Air Force (USAF) was documented in the companion report (Mark Toukan, Matthew Walsh, Ajay K. Kochhar, Emmi Yonekura, and David Schulker, *Air Force Operational Test and Training Infrastructure: Barriers to Full Implementation*, RAND Corporation, RR-A992-1, 2022).

performance, readiness-related Air Force Instructions (AFIs), Defense Readiness Reporting System–Strategic (DRRS-S) squadron reports, Joint Tactical Air (TACAIR) Synthetic Training Environment (JTSTE) technical documents, and Common Synthetic Training Environment (CSTE) draft design plan documents available on ongoing investments in synthetic training environments.

In the following sections, we summarize of each of the main approaches to research that we employed: (1) discussions with senior leaders, technical experts and SMEs; (2) review and analysis of current readiness reporting, and (3) analysis of ongoing development efforts related to new synthetic training environments with a supporting literature review.

Senior Leader and Subject-Matter Expert Discussions

We conducted semistructured discussions with Air Force senior leaders—identified with the assistance of the Air Force sponsor representative—from Air Combat Command (ACC), Air Mobility Command (AMC), and Air Force Global Strike Command (AFGSC).[9] With each senior leader, we had either one 60-minute discussion or two 30-minute ones, depending on availability. These discussions covered three main topics: (1) the working definition of *readiness* on which the senior leader relies in the decisionmaking process; (2) the utility of readiness metrics for decisionmaking; and (3) the role of live, virtual, constructive (LVC) assets for readiness assessment. The appendix presents the detailed protocol we used in these discussions. We acknowledge the limitation that our findings are based on interviews with four senior leaders that represent a subset of all MAJCOMs. There are potentially more readiness definitions and reporting issues than those identified in our report.

We identified SMEs and technical experts at the primary offices involved in readiness reporting, training policy, and training infrastructure at Headquarters Air Force and from ACC, AMC, AFGSC, and Air Force Special Operations Command. We then conducted nine interviews using a guide compiled for these discussions that included questions in three broad areas: current readiness reporting processes across the Air Force and unique to specific MAJCOMs; the logic of readiness and gaps in existing readiness metrics; and current capabilities to measure readiness and potential investments for building better metrics, including gathering, storing, and analyzing data from training infrastructure. Given the small number of interviews, we conducted a manual analysis of the transcripts and identified the major common themes present across the interviews.

Review and Analysis of Current Readiness Reporting

We analyzed a snapshot of DRRS-S reports for capability readiness for the month of November 2020, focusing on active-duty operational units.[10] We examined the squadron

---

[9] All senior leaders were either a general officer or a Senior Executive Service member.

[10] The units included in our analysis were all active-duty flying units that report readiness in DRRS-S. We excluded maintenance, logistics, and command-and-control units.

commander comments for each Mission Essential Task (MET) in a unit's capability assessment to determine what types of readiness information the comments contained that had not already been communicated through the resource and MET ratings. Each comment was labeled with primary and secondary themes and categorized as either a "unit-level" or "senior leadership–level" concern according to which authority would be most appropriate to address the issue. Last, we performed a quantitative analysis to determine the frequency of themes across platforms, MAJCOMs, and METs.

Analysis of Ongoing Development Efforts

With sponsor guidance, we looked at draft technical development plans for the JTSTE, a potential future component of LVC for OTTI. JTSTE comprises three separate components: the Joint Integrated Training Center (JITC); the Joint Simulation Environment (JSE) Linked; and the CSTE.

With further sponsor guidance on investment priorities, we narrowed the focus to CSTE. Given the vision for CSTE, we outlined the ways in which CSTE can help address readiness assessment gaps. The foreseeable technical challenges for CSTE were distilled from technical documents and discussions with SMEs, and each challenge is connected to implications for readiness and readiness assessment to be considered in the design decisionmaking process.

In addition to the specific look at CSTE, we also conducted a more-general literature review on developments in OTTI concepts and technologies that are relevant to readiness assessment. This literature review included academic, industry, and defense-related sources (both U.S. and international). In total, 94 documents were reviewed, with an emphasis on papers that specifically looked at the USAF and other military populations and on recent papers (i.e., past five years). The literature we reviewed came from several sources. Some was provided directly to us following technical SME discussions. Other literature came from searching in academic, defense, other government (e.g., Government Accountability Office), and industry (e.g., Interservice/Industry Training, Simulation and Education Conference) databases for such terms as *training technology*, *training tool*, *competency model*, *distributed mission training*, *adaptive training*, *training simulator*, and *team performance*. Additional literature was also sourced from recommendations from engaging with researchers who authored papers in our initial literature search.

Organization of This Report

We begin in Chapter 2 with a discussion of the different definitions of readiness and how the differences affect readiness assessment for Air Force senior leaders. In Chapter 3, we discuss measurement gaps in readiness assessment, where data are lacking to inform senior leader decisions. Next, in Chapter 4, we discuss ways that OTTI can address the readiness assessment gaps, starting with a discussion of the investments in OTTI that senior leaders think would improve readiness assessment. Then, in Chapter 5, we look at a specific training infrastructure investment and CSTE and discuss technological barriers and their implications for readiness assessment. The chapter closes with a review of other technologies that should be considered to address the readiness assessment gaps. The report proper closes in Chapter 6 with conclusions and recommendations. The appendix presents our protocol for the senior-leader interviews.

2. Defining Readiness

Ultimately, it is most important that senior leaders are getting the most operationally relevant readiness assessment information to support their decisions in the current strategic environment. However, ensuring the best outputs starts with examining the entire process. At a very basic level, readiness assessment involves defining *readiness*, defining its standards of evaluation, observing a demonstration of readiness, and evaluating whether the demonstration measures up to the defined standards (top row of Figure 2.1).

Figure 2.1. Readiness Assessment Process

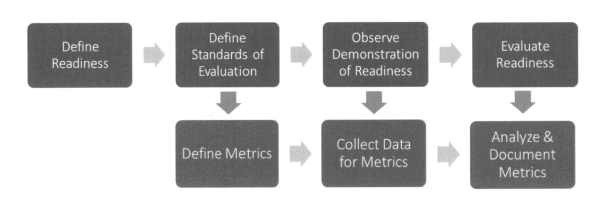

NOTE: The upper row of boxes represents the main readiness assessment process, and the second row indicates where metrics considerations come into play.

For an efficient assessment, the entire readiness assessment process should rely on a consistent definition of readiness. The definition affects how one will assess, demonstrate, and report readiness. Once a definition is established, the standards of evaluation can be defined to ensure all aspects of readiness are held to account. These standards can then guide what needs to be done to conduct a readiness demonstration. As shown in Figure 2.1 (bottom row), beginning the process by thoughtfully defining readiness allows definition of metrics that intentionally align with readiness inputs, activities, outputs, and/or outcomes. Similarly, if metrics are carefully predefined, data collection, analysis, and documentation can be integrated into the design of a readiness demonstration. Conversely, if readiness is left ill-defined, decisions made in later steps in readiness assessment could constrain the type of readiness that can be assessed.

In this chapter, we review definitions of readiness, show how current readiness reporting aligns with certain definitions, then explore perspectives from four Air Force senior leaders on readiness definitions and metrics.

Broadening the Definition of Readiness

The official DoD definition of *readiness* is "the ability of military forces to fight and meet the demands of assigned missions."[11] Others have pushed for a new definition that better reflects the priority of readiness for a high-end fight with a near-peer adversary.[12] A 2017 Congressional Research Service report contains an in-depth discussion of DoD readiness definitions and delineated both a *narrow* and a *broad* definition. The narrow definition is focused more on unit or operational readiness—on the capability of a unit to perform the missions for which it was organized or designed. In this way, the narrow view is only one readiness element, along with force structure and modernization, that determines the capability of a military force.

Definitions of Readiness

Official DoD definition of *readiness*: "the ability of military forces to fight and meet the demands of assigned missions" (JP-1, 2017)

***Operational readiness* (the narrow interpretation):** the capability of a unit to perform the missions for which it was organized and designed (Rumbaugh, 2017)

***Broad readiness* (the broad interpretation):** the ability of all integrated military forces to conduct the full spectrum of operations against any adversary.

The broad definition of readiness aligns with the official DoD definition to consider the bigger picture of integrated military forces and their ability to conduct the full spectrum of operations against any adversary.[13] A subsequent Congressional Research Service report carefully parses the official DoD definition to develop a readiness framework with three levels of readiness assessment:

1. the subunit level, with the different aspects of readiness that contribute to the unit (e.g., medical readiness)
2. the unit level
3. all military forces.

The third readiness assessment is most aligned with the broad definition of readiness. The other two levels of readiness assessment are essentially thought of as readiness inputs, but they also align with the narrow definition of readiness.[14] More recently, the discourse over defining

---

[11] Joint Publication (JP) 1, *Doctrine for the Armed Forces of the United States*, Joint Chiefs of Staff, incorporating change 1, July 12, 2017, p. GL-10.

[12] The 2018 NDS articulates the top priority as long-term strategic competition with near-peer nations and also that the joint force should be able to prevail over aggressions from a major power (Mattis, 2018).

[13] Russell Rumbaugh, *Defining Readiness: Background and Issues for Congress*, Congressional Research Service, R44867, June 14, 2017.

[14] G. James Herrera, *The Fundamentals of Military Readiness*, Congressional Research Service, R46559, October 2, 2020.

readiness continued with a framework advocated by the Chief of Staff of the Air Force and the Commandant of the U.S. Marine Corps. They proposed a broad readiness framework that includes current availability of personnel and equipment; functions across combatant commands; and future availability, readiness, and modernization efforts. The framework would incorporate risk assessments that consider risks to specific force elements, duration, and probability. Underlying the framework is the concern that DoD legacy weapon systems will never be sufficiently capable for a near-peer adversary conflict; thus, looking toward future readiness and modernization is key. Using the proposed framework, complex trade-offs could be made on a global scale over space and time.[15]

The framework proposed by the two service leaders also acknowledges the military readiness framework proposed earlier by Richard Betts, which centers on three questions:

1. How much capability is needed to achieve objectives in wartime scenarios? ("Ready for what?")
2. How much time will be available to convert assets into capabilities, e.g., by mobilizing reserve units? ("Ready for when?")
3. How should the different elements that combine to form capabilities be managed? ("Readiness of what?").[16]

In the Betts framework, readiness would be defined for a specified set of forces, time frame, and operational scenario. A *ready force* is one that can supply all needed capabilities to meet wartime demands; an *unready force* is one that lacks the ability to meet the demand for capabilities. A simplified way of thinking about assessing readiness for this definition would be to evaluate the expected performance of forces against potential scenarios, with time horizons and production capacity included as scenario dimensions.

The important readiness questions that senior leaders are asking today—such as, can joint forces operate in a cyber- or space-contested environment?—fit best with a broader definition of readiness and, thus, a broader readiness assessment. In response to policy questions that exceed the scope of traditional readiness assessments, policymakers have moved to bring new dimensions to the definition of strategic readiness (of which narrow operational readiness is one component).[17] However, these discussions on the need for a broader definition of readiness have not yet produced fundamental shifts in the readiness assessment and reporting process, as we discuss in the next section.

---

[15] Brown and Berger, 2021.

[16] Betts, 1995.

[17] Bradley Martin, Michael E. Linick, Laura Fraade-Blanar, Jacqueline Gardner Burns, Christy Foran, Krista Romita Grocholski, Katherine C. Hastings, Sydney Jean Litterer, Kristin F. Lynch, and Jared Mondschein, *Measuring Strategic Readiness: Identifying Metrics for Core Dimensions*, RAND Corporation, RR-A453-1, 2021.

Current Readiness Reporting and Readiness Definitions

As part of DoD's official readiness reporting system, DRRS,[18] the DAF reports on two different types of readiness measures at the unit level:

- *Resource readiness* indicates that a unit possesses the required resources and is trained to undertake the full wartime mission(s) for which it is organized or designed. Its reporting is meant to be an objective assessment of unit operational readiness and a tracking mechanism for organizing, training, and equipping forces for combatant commands.[19]
- *Capability readiness* indicates the ability of an organization to accomplish its METs to the established standards based on the organization's capabilities, under conditions specified in its joint or agency MET list. It is reported from the unit commander's subjective assessment of whether the unit's trained personnel and assigned equipment can conduct the METs that embody the capabilities the unit was designed to execute.[20]

Some important differences between the two readiness measures include the fact that resource readiness does not consider different threat environments.[21] Capability readiness is informed by objective standard assessments within each MET, although the overall rating of capability assessment is ultimately subjective. To provide historical context for why there are two types of readiness measures:

> In 1999, the DoD began developing the Defense Readiness Reporting System (DRRS) in response to criticisms and shortfalls in the SORTS [resource readiness reporting]. The DRRS [capability readiness reporting] was initially intended to replace the SORTS but has since been modified to include SORTS metrics and improve upon the SORTS reporting system. Perhaps the most significant difference with the DRRS is the inclusion of a commander's self-assessment of whether a unit is ready to perform the missions and tasks assigned to it on a three-level scale: yes, qualified yes, and no.[22]

One of the purposes of the transition to DRRS and inclusion of capability readiness was to emphasize the measurement of *readiness outputs* instead of solely focusing readiness measurements on *resource inputs*. Thus, the capability assessments guard against the possibility that a fully resourced unit might appear ready even if its design is fundamentally inadequate for its mission. It is noteworthy that the data in DRRS are used across echelons for myriad purposes.

---

[18] The system the DAF uses is called *DRRS-Strategic* (DRRS-S), however, we use *DRRS* in this chapter because it is commonly referred to as such.

[19] This was formerly tracked in the Status of Resources and Training System (SORTS) and includes ratings for personnel, equipment on hand, equipment status, and training levels (i.e., P-level, S-level, R-level, and T-level, respectively). The worst rating of the four determines the overall unit resource rating, called the C-level. They are also commonly referred to as P-rating, S-rating, etc.

[20] AFI 10-201, *Force Readiness Reporting*, December 22, 2020.

[21] An exception is the training piece of resource readiness. Training requirements include events in different threat environments; however, no specific metric tracks ability to train in, or performance in, any threat environment.

[22] Todd Harrison, "Rethinking Readiness," *Strategic Studies Quarterly*, Vol. 8, No. 3, Fall 2014.

At the unit level, commanders may use readiness assessments to guide unit management (e.g., spend more time training) and to communicate resource needs up the chain of command. At higher echelons (i.e., wing, MAJCOM), decisionmakers draw on unit capability assessments to understand what can be presented as capabilities to combatant commands and use resource assessments to inform tasking and resource needs. At even higher echelons (e.g., Joint Staff), leaders use further aggregated assessments to understand whether forces are adequate to meet National Military Strategy objectives. The multiple different uses of DRRS make it difficult for the data to meet all needs as intended.

Two decades since the conception of DRRS, the two types of reporting have persisted but not without friction. For example, in examining DRRS data, we found examples of misalignment between resources and capability ratings. These misalignments occur when a unit will rate as fully resourced but have a MET that is not rated as "Yes," or when MET(s) are rated as "Yes" but a resource is not at its top rating. Such misalignments are explained in the commander remarks and are included in the analyses shown in MAJCOM-level readiness briefings.

One explanation for misalignments is the perception that resource readiness captures the expectations of the unit as it was organized, trained, and equipped, while capability readiness includes future aspirational expectations, such as operating in certain threat environments. The reverse case, in which units are underresourced yet capable, could indicate that resourcing standards are overly conservative or that unmeasured aspects of the inputs are able to compensate for the resource shortfalls. In that sense, resource readiness measures the conformity with the designed force structure and flags mismatches for senior leader attention. Capability readiness, on the other hand, uses subjective judgment to determine whether resources will translate into mission accomplishment, where missions span different conditions and environments.

In terms of alignment with narrow or broad definitions of readiness discussed in the previous section, both resource and capability readiness have the flexibility to align with either. In practice, because most assessments occur within squadrons, current readiness reporting is primarily aligned with a narrow definition (operational readiness). Capability readiness spans definitions in that each MET is evaluated against different conditions and threat environments, and some METs are also tied to operational plans (OPLANs).[23] However, getting the full perspective on military forces requires rolling up all units that report to a MET or all reported METs for an OPLAN.

Current reporting also considers the "ready for when?" question with the inclusion of unit response time, which is specified by the MAJCOM; units report whether their resources will be ready and available within the designated response time. OPLANs may also specify a different

---

[23] OPLANs are detailed plans for joint military operations developed by combatant commanders, and they are in the context of real or potential military situations.

response time.[24] Despite current practices, resource and capability readiness could (in theory) match up with a broad definition of readiness if it were supported by DAF policy and documentation that specified well-defined potential threats and time frames that look to the future.

Another key issue in current readiness reporting is the "readiness of what?" aspect: DRRS tracks readiness at the unit level, typically by squadron; however, the unit of force presentation is not the squadron. Instead, the DAF presents forces in terms of Unit Type Codes (UTCs). A UTC is a potential capability, represented by a package of equipment or personnel, that is focused on accomplishment of a specific mission that the military service provides. Units are responsible for producing specific UTCs, and unit commanders report on their UTC's ability to perform their mission across the full range of military using the Air and Space Expeditionary Force UTC Reporting Tool (ART), outside DRRS.[25] UTC readiness reporting supports global force management. Considering both DRRS and ART readiness reporting, Figure 2.2 captures the different pieces of resource and capability readiness at various organizational levels, where only the unit resource readiness is objectively assessed.

Figure 2.2. Current Readiness Assessment

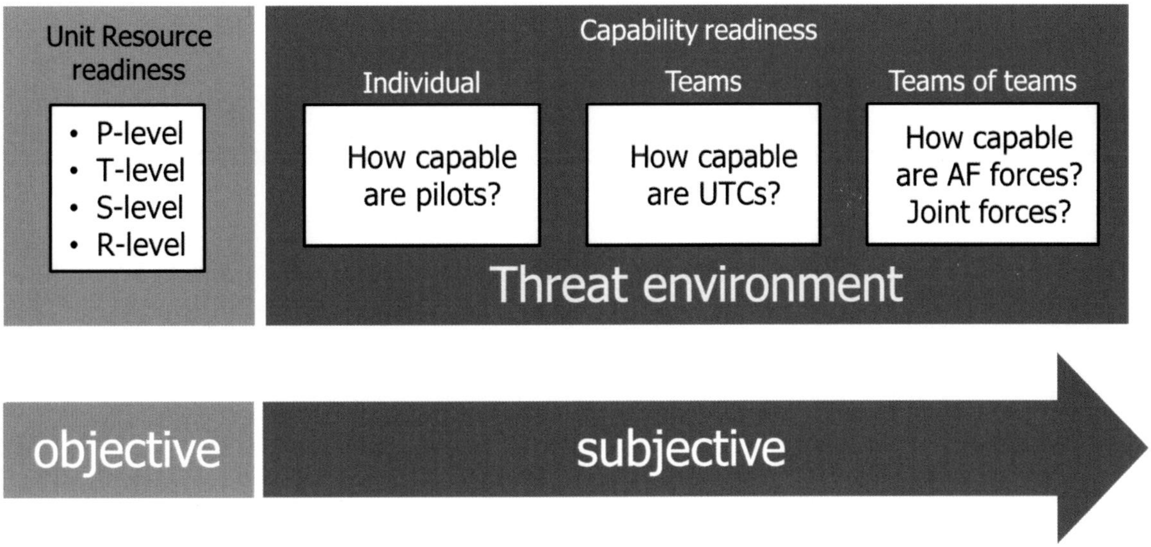

SOURCE: RAND analysis of AFI 10-201, 2020.

Readiness Definitions and Metrics in Practice

In our interviews with four senior leaders, we learned that the legacy readiness measures (rather than a specific need) drive how the leaders define readiness and use readiness data.

[24] AFI 10-201, 2020.

[25] AFI 10-244, *Reporting Status of Air and Space Expeditionary Forces*, U.S. Air Force, June 15, 2012. Note that the ART tool is being replaced by the Deliberate Crisis Planning and Execution Segments application.

11

During our discussions we noticed that the operational definition of readiness on which they rely is aligned with DoD's doctrinal definitions of resource and mission capability readiness,[26] which supports our assessment that the legacy readiness reporting system aligns most with the narrow definition of readiness.

We also noticed during these discussions that these senior leaders showed a preference for readiness definitions that focused mainly on resources. For example, one senior leader mentioned that the definition of readiness they used focused on whether the materiel needed to carry out the required task were available to the crew in due time,[27] while another senior leader pointed to the need for units of action (i.e., squadrons, individual crew members) to have the resources and training they need to execute the METs that they have been assigned.[28] These operational definitions of readiness are aligned with the doctrinal definitions of resource readiness and capability readiness.[29]

These senior leaders mentioned that they preferred resource readiness metrics because they are more useful to the leaders than capability readiness metrics.[30] During our discussions, senior leaders stated that resource metrics are easy to understand and make it easier for decisionmakers to identify resource shortfalls and, in turn, identify where additional resources need to be redirected.[31] In this vein, one of the Air Force leaders with whom we spoke outlined the benefits of resource readiness metrics (e.g., C-level): They are quantifiable and make it easy for leadership and decisionmakers to evaluate "what a unit does or does not have [in terms of resources] and at what level they are able to train."[32]

Despite these senior leaders' common preference for resource metrics, our qualitative analysis that identified and manually coded the major themes present across the transcripts of the discussions showed that views among MAJCOM leadership diverged regarding the value of various binary metrics (i.e., supply rates). While some leaders would find binary metrics to be generally useful "because they are less interpretative," they might find specific ones, such as T-ratings, to obscure the "ready for what?" question: The underlying training requirements have changed over time from a focus on counterinsurgency to high-end conflict.[33] At the same time, for other senior leaders, T-ratings speak to some extent to scenario performance,[34] or they need a

---

[26] U. S. Government Accountability Office, *Department of Defense Domain Readiness Varied from Fiscal Year 2017 Through Fiscal Year 2019*, GAO-21-279, 2021.

[27] Discussion with Air Force official, May 17, 2021.

[28] Discussion with Air Force official, April 27, 2021.

[29] See definitions for resource readiness and capability readiness earlier in this chapter.

[30] Discussions with Air Force official, April 27, 2021, and May 14, 2021.

[31] Discussion with Air Force official, April 27, 2021.

[32] Discussion with Air Force official, April 27, 2021.

[33] Discussion with Air Force official, May 14, 2021.

[34] Discussion with Air Force official, April 27, 2021.

more nuanced interpretation because a "T-rating might make [one] unit look healthy, but [a] specialized unit not so much."[35] Furthermore, when it comes to P and T ratings, one senior leader mentioned an internal Air Force assessment that showed that these two metrics were the two most important drivers of resource readiness.[36] Another senior leader considered the tension between T- and P-rating data and that "this tension is particularly productive" (i.e., "how to schedule [training] exercises to get the most opportunity") with future analytics ideally being able to capture this tension.

Senior leaders pointed to capability assessments as less or the least useful readiness metrics. While qualitative DRRS assessments are needed to assess force readiness in the absence of data aggregation, capability assessments are perceived to be less useful than resource metrics because the former are subjective and require decisionmakers to dig deeper to understand on what the metrics are based.[37] With decisionmakers often facing a fast-paced decision environment, it is not always possible (and realistic to expect) that they have the time and bandwidth to carry out an in-depth analysis to understand what drives the respective qualitative assessments—constraints that are especially problematic for large MAJCOMs. However, in smaller MAJCOMs, such as AFGSC, senior leaders have the opportunity to discuss the issues at hand with squadron commanders and understand what is behind the data on capability assessments.[38]

Furthermore, one senior leader mentioned the challenge of identifying trends in data across multiple units using capability assessments, which subsequently makes it difficult to ensure that commanders have the resources that they need to execute their missions. As a result, senior leaders rely on C-level data more often than on capability assessments, and so does the Joint Staff, which does not see qualitative assessments to be very useful for contingency planning. In this light, when discrepancies occur between C-level reporting and qualitative assessments—for instance, when squadron-level C-ratings show that resources are short, but the qualitative assessment says that they "are ready to go"—commanders usually turn to C-levels because they are easier to measure.[39] However, one senior leader mentioned the importance of discussing qualitative ratings with wing commanders, who assess whether a certain unit is "good to go in spite of C-rating."[40]

Despite the senior leaders' preference for C-level metrics, the overall assessment is that DRRS has limitations and that legacy readiness metrics are not well suited for the missions that

---

[35] Discussion with Air Force official, May 17, 2021.

[36] Discussion with Air Force official, April 27, 2021.

[37] Discussion with Air Force official, April 27, 2021.

[38] Discussion with Air Force official, May 14, 2021.

[39] Discussion with Air Force official, April 27, 2021.

[40] Discussion with Air Force official, May 17, 2021.

the NDS threats would require.[41] As one senior leader pertinently remarked, "DRRS [is] only as good as what you are grading it against."[42] The same senior leader also noted that the legacy readiness metrics are very much focused on the individual (i.e., individual training for individual missions), while most NDS missions require an integrated approach, which means that the "what and how of training," and—implicitly—of readiness reporting, have to change.[43]

Implication

The readiness assessment process is unique for each definition of readiness. While readiness reporting inputs could theoretically inform broader questions, our discussions with a few Air Force senior leaders indicated that current practice focuses on using metrics to make sure units are properly resourced so that they can be tasked. A limited assessment at the unit level, based on a limited definition, may be acceptable for determining short-term tasking for MAJCOMs. However, a broader definition of readiness (e.g., Betts) is needed to provide senior leaders with readiness information that elucidates strategic shortfalls. The broader definition should answer the three questions that the Betts readiness framework poses by specifying such things as mission type, mission conditions, integration with other units, and specific adversary capabilities and tactics. In Chapter 3, we discuss how the design of the reporting system creates shortfalls in the ability of readiness assessments to answer questions regarding strategic capabilities of forces.

---

[41] This limitation is also reflected in our analysis of the DRRS-S commander remarks associated with METs: The majority of comments focus on unit-level resource issues rather than on higher-level concerns of modernization and force structure.

[42] Discussion with Air Force official, June 11, 2021.

[43] Discussion with Air Force official, June 11, 2021. However, current Ready Aircrew Program requirements include training events above the individual level.

3. Gaps in Readiness Assessment

Aside from the issue of defining readiness, as discussed in Chapter 2, the Air Force is required to assess and report readiness. We examined the readiness assessment process for any existing knowledge or data gaps. In our analysis of existing readiness metrics and our discussions with senior leaders and SMEs, three major themes emerged around the topic of readiness assessment gaps:

1. gaps associated with factors present when forces are integrated
2. gaps associated with aggregated readiness reporting that does not match force presentation
3. gaps in readiness assessment associated with scenarios for and threats against which forces cannot (or rarely) train.

In this chapter, we will present these three main themes along with supporting evidence for each. Before getting into a discussion of the themes, we will provide some background on readiness reporting.

In current practice, squadrons report on their unit resources and capabilities and the UTCs for which they are responsible.[44] Unit readiness includes the resource (i.e., P, S, R, and T representing personnel, equipment on hand, equipment status, and training, respectively) and capability (i.e., Y for "Yes," Q for "Qualified Yes," and N for "No") ratings. The capability ratings take the form of multiple METs that cover capabilities that the unit was designed to execute. Each MET includes a set of conditions—such as day or night, low or high altitude—and measures, which catalog whether specific resources meet a specified level, whether a subcapability exists, and whether the task can be conducted under a specified threat environment. Given the current values of the measures, the unit commander makes a judgment as to whether the unit can perform the MET. UTCs represent packages of personnel or equipment that would be deployed and are meant to represent a unit of capability (e.g., six aircraft). Multiple UTCs can make up a force package that includes the operations, maintenance, and munitions capabilities required for a mission. Together, and potentially with other force packages, they must execute a mission in a specific operational scenario. UTC readiness is assessed by the unit commander, who assigns a rating of Y, Q, or N to each UTC.

Gap 1: Factors That Come into Play Only When Forces Are Integrated

Capabilities, such as suppression of enemy air defenses or close air support, are not delivered by one individual but by an integrated team. However, integration is not well measured within or

---

[44] AFI 10-201, 2020.

across units, UTCs, or force packages. The current readiness assessment process does not contain inputs concerning factors that cross unit boundaries and factors that affect the integration of building blocks.[45]

Figure 3.1 illustrates the different instances of force integration and where readiness assessment takes place. Starting with the unit (i.e., squadron), for which resource assessments (represented by the boxes with P, S, R, and T) and capability assessments (represented by the boxes with different METs) are conducted. Units are responsible for generating the personnel (P) and/or equipment (E) that make up different UTCs, which represent subunits used for force presentation and can represent such capabilities as operations (ops), maintenance (mx), and munitions (mun). UTCs combine to make force packages, which can work together with other force packages during an operational scenario. At the top of the figure is the joint task force, which acknowledges that DAF force packages may team with other joint forces in a particular operational scenario.

Figure 3.1. Readiness Assessment and Levels of Integration

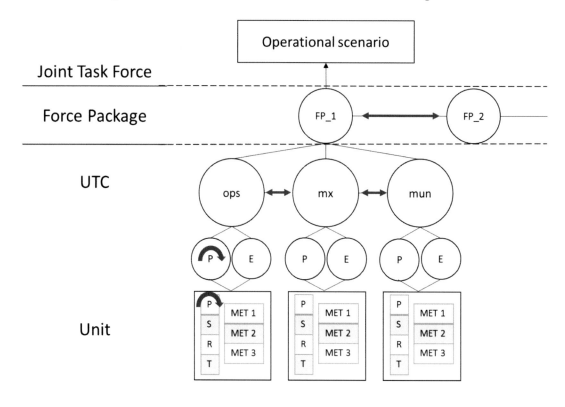

NOTE: USAF units (i.e., squadrons) are responsible for producing and reporting on multiple UTCs.

---

[45] Factors of integration may be captured if a unit has recently engaged in an exercise, which can be cited in a commander's assessment. However, exercises do not occur frequently enough to inform monthly assessments across all personnel in all units. Also, a unit commander must implicitly consider integration when rating the unit's ability to perform its METs and present UTCs, which may be based on observations of training sorties.

In Figure 3.1, the red arrows indicate example areas where integration factors come into play but are not captured by current readiness reporting. Starting at the bottom of the figure,

- The rounded arrows in the unit and UTC represent how the readiness levels of individual personnel are aggregated to create the readiness measurement for the unit or UTC without necessarily assessing how teams of personnel integrate and work together.[46]
- Individual UTCs must be integrated with other UTCs to provide capability to a force package. The red arrows between ops, mx, and mun represent how three types of UTCs must integrate to perform their functions as a force package; however, the readiness of the force package is taken as the aggregated readiness from individual UTC types.[47]
- Similarly at the force-package level, individual packages must be integrated with one another to meet the demands of an operational scenario.

To the extent that unit commanders capture integration in assessing their units, this unit-level integration does not necessarily translate into integration of UTCs and force packages. More generally, only the unit level is measured, and the assessment system does not map levels horizontally or vertically to one another. As a result, readiness assessments of force packages or joint task forces may not attribute a lack of readiness with sufficient specificity so that decisionmakers can invest in the appropriate places. In looking for the root cause of a readiness issue, a senior leader can drill down to unit or UTC readiness ratings, but the readiness system will not flag an issue, for example, related to coordination between two UTCs. These missing factors mean that senior leaders do not understand where to invest to buy the most resource or capability readiness, e.g., investments in operations and maintenance of current forces versus decisions that shape long-term strategy, acquisition, and force structure.

We have seen evidence of this gap across the different levels of our analysis. A senior leader noted that the readiness assessment system does not capture the ability of units across the mobility and combat air forces to integrate for a high-end mission.[48] Simply aggregating the set of existing unit-level metrics—e.g., the readiness of a fighter unit on the one hand and a refueling unit on the other—does not translate meaningfully into the readiness of an aggregated combat capability. The fulfillment of training requirements by each type of unit only implicitly demands some degree of integrated readiness. The current system, which takes squadrons as the foundational measured unit, produces metrics that can only extrapolate the speed and capability of UTCs to deploy together and generate missions, as one senior leader noted.[49] Another senior leader explained that the scarcity of complex, integrated training environments limits the ability to identify where more or different types of integration, for example, between air and ground

---

[46] AFI 10-201, 2020; AFI 10-244, *Reporting Status of Air and Space Expeditionary Forces*, supplement, U.S. Air Forces in Europe, December 21, 2017.

[47] Typically, MAJCOMS use such tools as the Posturing Analysis Tool or the Deliberate and Crisis Action Planning and Execution Segments system to conduct this assessment.

[48] Discussion with Air Force official, May 5, 2021.

[49] Discussions with Air Force official, April 27, 2021, and May 15, 2021.

forces, might be necessary to increase readiness against specific threats and operational scenarios.[50] The latter point underscores that the issue goes beyond the DAF.

The DAF recognizes that these points of integration between capabilities will become more important in the future than they have been in the past, which is reflected in the DAF pursuit of new operational concepts for integrating capabilities (such as Joint All-Domain Command and Control, as well as Agile Combat Employment). To illustrate, consider METs for performing close air support. In counterinsurgency operations in the Middle East and Central Asia, the lack of an enemy integrated air defense system meant that readiness for close air support METs was mainly a question of individual pilot targeting and weapon employment proficiency, along with the status of his or her platform and munitions (although targeting would involve some integration with intelligence, surveillance, and reconnaissance and with forward air controllers).

In contrast, understanding unit readiness for close air support in a high-end conflict involves many more factors related to integration because this mission would involve a network of intelligence, surveillance, and reconnaissance platforms to determine the status of enemy air defenses, companion capabilities (such as offensive counterair) to respond to threats, electronic warfare platforms to degrade the effectiveness of enemy defenses, increased coordination with larger ground-force elements, and command-and-control functionality to pass information across the environment. Thus, whether a unit or UTC can accomplish close air support METs simply captures the piece of the mission that the particular unit contributes and practices. However, the implicit and unproven assumption is that, if all the pieces can do their part, they will work successfully as a whole.

Gap 2: Readiness Report Aggregation Does Not Match Force Presentation

Readiness reports are currently aggregated in a way that does not match DAF force presentation. There is some similarity between our Gaps 1 and 2; however, the Gap 2 relates more to a design flaw in the readiness reporting system. It is only at the unit level that capabilities are assessed for a range of different environments and threat levels, which is the type of information that relates to different operational scenarios (illustrated in Figure 3.2 with blue shading). Key information on capabilities under specific conditions and in various threat environments is extrapolated from the unit level to the UTCs, force packages, and joint task force.[51] A readiness system closer to the ideal would allow a senior leader to track the readiness of operational scenarios by looking at the forces that would actually be deployed.

---

[50] Discussion with Air Force official, May 14, 2021.

[51] Mane and colleagues have provided recommendations for how to expand the use of UTCs to be more useful for decisionmakers at ACC to assess readiness with respect to different scenarios (Muharrem Mane, Anthony D. Rosello, Paul Emslie, Thomas Edward Goode, Henry Hargrove, and Tucker Reese, *Developing Operationally Relevant Metrics for Measuring and Tracking Readiness in the U.S. Air Force*, RAND Corporation, RR-A315-1, 2020).

Figure 3.2. Gap 2: Readiness Report Aggregation

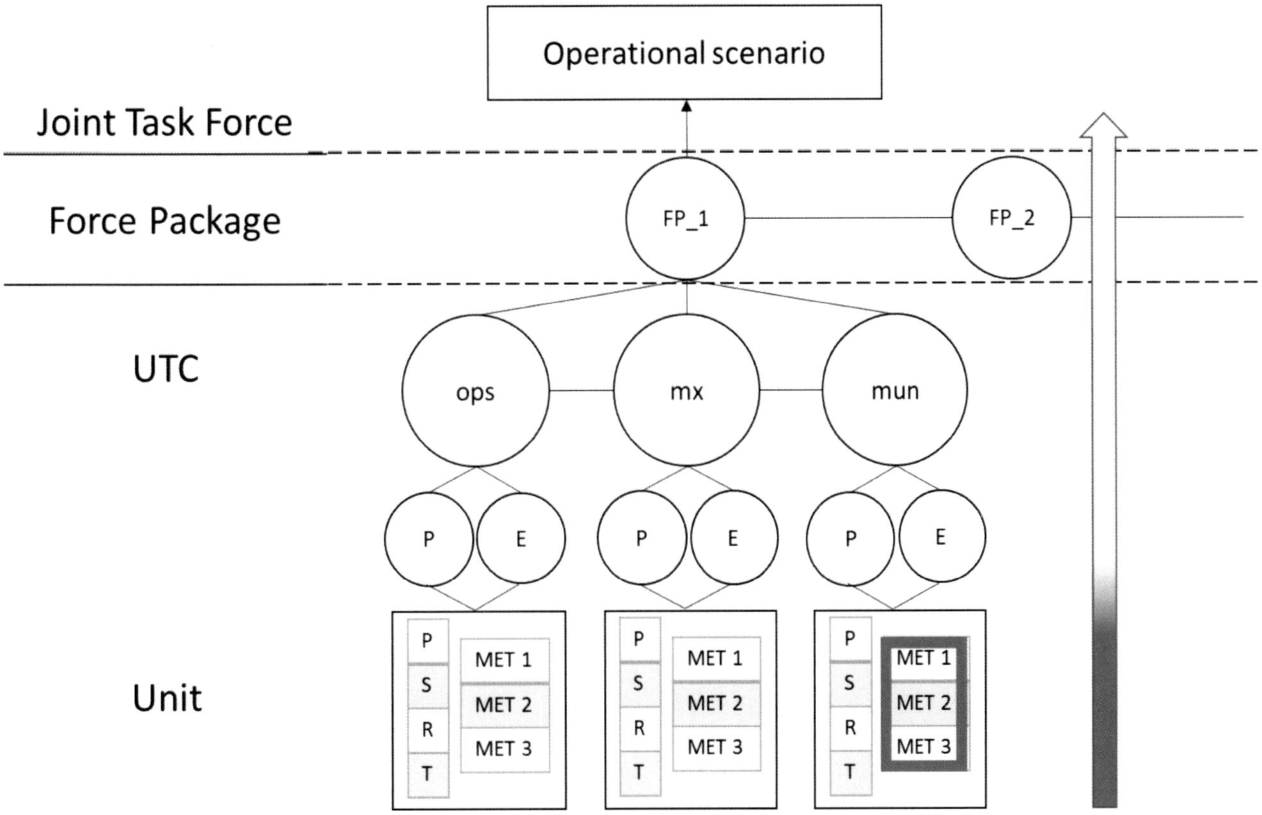

NOTE: The shaded arrow on the right indicates where information on capabilities in a threat environment is concentrated and then extrapolated.

One senior leader stated that current readiness aggregation presents a big shortfall because the Air Force is evolving to view wings as the aggregated fighting unit.[52] The same senior leader gave an example of how readiness aggregation does not work for the many squadrons that have associated Guard or Reserve units. Even though these units may be operating the same weapon system, they report readiness separately up to two different commands, which are essentially disparate systems.[53]

One potential mitigation for the lack of aggregated readiness assessments at higher levels and the previous readiness assessment gap is to conduct large-scale exercises to assess force integration and readiness in the context of different operational scenarios.[54] This type of readiness assessment would then need to be documented in a new expansion of the existing

---

[52] Wing commanders are already responsible for reviewing the readiness of their subordinate units and may also have observed evidence of integration between different functions at the base.

[53] Discussion with Air Force official, May 19, 2021.

[54] One way this was accomplished in the past was through cyclical, multiunit operational readiness inspections run by the MAJCOM Inspector General.

readiness reporting system that speaks to force packages instead of units. However, such exercises are expensive, too infrequent to cycle through all DAF units and all desired operational scenarios, and limited by such factors as operations security. These limitations associated with conducting large-scale exercises virtually are discussed as part of the next readiness assessment gap. As a result of this readiness aggregation gap, senior leaders must work with extrapolated and subjective information when assessing force readiness in high-end threat environments.

Gap 3: Commanders Report Readiness on Threats and Scenarios They Cannot or Rarely Train Against

Opportunities can be nonexistent or scarce for units to practice and demonstrate proficiency for certain capabilities across all required environments in their readiness reports. Lacking sufficient data, assumptions about unit readiness are extrapolated from limited information, or unit readiness is simply not reported due to lack of information. This gap in readiness assessment has direct ties to the available training infrastructure.

An implication of this gap, similar to the previous gap, is that the readiness data senior leaders need for what are, arguably, the most important scenarios either do not exist or are based on a highly subjective commander assessment. The inability to train for certain threats or scenarios has implications for assessing force structure and for the ability to generate force packages with specific capabilities over time to succeed under a range of operational scenarios.

This gap and its consequences are supported by our discussions and DRRS-S analysis across MAJCOMs. One senior leader stated that, while a fifth-generation squadron might be 85 percent combat mission ready, how it will perform against high-end threats must be based on a subjective assessment. In support of that point and to generalize, readiness SMEs across MAJCOMs explained that training against high-end threats is rarely available at home station.[55] Such training is primarily encountered at large-scale exercises, such as Red Flag. [56] The SMEs mentioned that, even at Red Flag, the capability to assess performance levels against certain threats and record them in the readiness system is lacking.[57] The relative scarcity of opportunities to train against such threats and the lack of tools to capture performance data mean that commanders must make subjective, incomplete assessments of the capability of their units in the most important operational scenarios.

Synthetic training environments offer an alternative to live training, which, in theory, should fill in some of the training and assessment gaps currently experienced across the Air Force.

---

[55] Headquarters Air Force requires all Air Force operational units to report on METs at different threat levels that are associated with different threat environments. However, individual MAJCOMs are developing guidance on how to assess to these environments (discussion with Air Force officials, January 29, 2021).

[56] Discussion with Air Force readiness SMEs, March 22, 2021.

[57] Discussion with Air Force readiness SMEs, March 22, 2021.

However, these tools also have shortcomings. The quality, availability, and infrastructure to provide simulated threats differs across and within MAJCOMs. According to one readiness SME, threat representation is inadequate *when* the Distributed Mission Operations Network is working—when trainees can connect with other training systems across the force. An SME from another MAJCOM talked about the difficulties of setting up the same threat scenario across multiple devices because different platforms used in their MAJCOM have different threat representation systems.[58]

SMEs from yet another MAJCOM explained that training against high-end threats is only possible in the context of distributed mission operations (DMO),[59] but that DMO-connected training systems are in limited supply; furthermore, they require the resources of another MAJCOM to adequately simulate training in complex environments but do not have the funding or facilities to do so.[60] These observations are supported by our analysis of DRRS-S reports across MAJCOMs: Unit commanders do remark in MET comments whether there is an inability to train for specific threat levels.

Traditional readiness metrics in DRRS-S do not explicitly track the type and quality of training on which capability assessments are based. For instance, is the information based on a recent exercise or simulator training, or is it extrapolated from training in a benign environment? This information can be found as part of the commander's remarks and requires textual analysis to extract a metric that is actionable and can be analyzed across the force. The information may also be partially inferred by looking at the underlying training requirements (e.g., Ready Aircrew Program Tasking Memoranda), which specify frequencies of live or simulator events. However, the point remains that the current metrics obscure the quality of information. What is needed is a metric that specifically tracks the extent of Gap 3 such that senior leaders would be able to use it to track progress and the value of investments in new synthetic environments.

Implication

We have defined and characterized three gaps in the readiness assessment process and considered the implications of these gaps: (1) not accounting for factors that come into play only when forces are integrated, (2) readiness report aggregation that does not match force presentation, and (3) the fact that unit commanders must report readiness on threats and scenarios for which they either cannot or rarely train against. Addressing these gaps is not a simple matter of adjusting the current training infrastructure. Qualitatively different capabilities are needed to scale, integrate, and present complex scenarios and environments, which could be

---

[58] Discussions with readiness SMEs, February 16, 2021, and March 22, 2021.

[59] DMO is a type of training that leverages networked simulators to allow many warfighters across the services to rehearse missions in a synthetic operational environment.

[60] Discussion with readiness SMEs, January 29, 2021.

scheduled across units to aggregate force packages and executed to align with readiness reporting cycles. Furthermore, to fully close the gap in readiness assessment, the capabilities must allow some form of data collection to capture necessary and interpretable readiness measurements. In Chapters 4 and 5, we will discuss investments in training infrastructure that could address these gaps and particular training infrastructure design decisions that should consider the value-added to readiness assessment.

4. Perceptions of How Training Infrastructure Can Close Readiness Assessment Gaps

New synthetic training environments can address existing gaps in readiness assessment and better inform strategic decisions through two pathways. In the first, new training environments can improve the ability to rehearse operational scenarios (left side of Figure 4.1). More rehearsal obviously provides more practice to improve skills and more observations to inform readiness assessments, although the assessments may remain subjective. In turn, the improved capability assessments will provide senior leaders with better information on which to base strategic decisions. As we will discuss in this chapter, this pathway is prioritized because it *increases readiness* in addition to improving readiness assessment. In the other path, new synthetic training environments can also provide opportunities to collect and, potentially, to automate the collection of readiness-relevant data that objectively capture individual and team proficiency (right side of Figure 4.1). The stream of new data collected directly from the training environment can provide more-objective readiness metrics to enrich capability assessments. While the second pathway is less of a priority, it is key to making readiness metrics more objective and improving the data on which readiness assessments are based.

Figure 4.1. Pathways of OTTI Improvements

NOTE: CC = commander.

This chapter proceeds with perspectives from Air Force senior leaders on investments that the DAF would need to make in new synthetic environments to address the gaps in readiness assessment identified in the previous chapter. When asked about what LVC investments would help address the gaps, senior leaders identified five main investment areas: (1) DMO training;

(2) more simulators; (3) simulated threat environments; (4) measurement of aggregated readiness; and (5) adaptive, individual proficiency-driven training (see Table 4.1).[61] We connected these investment areas to relevant readiness assessment gaps.

Table 4.1. Senior Leader Perspective on LVC Investments

MAJCOM	Identified LVC Investments	Relevant Readiness Assessment Gaps
AMC, ACC, AFGSC	DMO training	Integration, aggregation, scenario
AMC, AFGSC	More simulators	Scenario
AMC, ACC	Synthetic threat environments	Scenario
ACC	Measuring aggregated force readiness	Integration, aggregation
AFGSC	Adaptive, proficiency-driven training	Integration, aggregation

SOURCE: Based on discussions with Air Force officials at different MAJCOMs.
NOTE: The relevant readiness assessment gaps refer to those discussed in Chapter 3 as follows:
Gap 1 = Integration; Gap 2 = Aggregation; Gap 3 = Scenario.

As Table 4.1 shows, the discussions with senior leaders provided a very high-level perspective on the LVC-related investments—essentially, training capabilities—that are needed to address existing gaps in readiness assessment. Most notably, the investment in DMO training was mentioned from several different angles, including a need for expanded and standardized simulated training environments that include more platforms and an IT infrastructure to support connectivity between simulators for different platforms. The remainder of this chapter discusses the five investment areas in further detail.

Investments in DMO Training

Investing in DMO training has many benefits for readiness, according to senior leaders across MAJCOMs. Three different types of technical investments would enable DMO training: expansion of training in a simulator environment, standardization of the simulator and training environment, and modernization of information technology (IT) infrastructure to support connectivity. Investments in each of these areas would contribute to closing all three readiness assessment gaps delineated in Chapter 3.

---

[61] As noted in our discussion protocol (see appendix), we specifically asked senior leaders about LVC investments. However, we left leeway for the multiple interpretations of *LVC*. The traditional definition would be a system that includes live, virtual, and constructive elements, but another common interpretation is to think of synthetic training systems that include only virtual and constructive elements.

Expand Training in a Simulator Environment to More Training Communities

At AMC, we learned of the need to expand training in a simulator environment, including DMO training for joint operations in which "multiple Blue assets train together in sim."[62] Currently, the availability of training in a simulator environment seems to be limited in all the services.

Training in a simulator environment could be expanded to elements of the joint force that currently train in a live environment that does not accurately reflect the threat level that the operators are likely to encounter in a combat situation. A simulator environment is more likely to deliver a threat emulation level closer to real combat conditions than what, for instance, 101st Airborne Division paratroopers would face during regular training jumps "into mountain ranges out West."[63]

While the expansion to a wider set of platforms and communities training in one simulated environment could potentially benefit the joint operations of several Air Force MAJCOMs, AMC—which, by the nature of its missions serves all the other MAJCOMs and supports training for the other armed services—would benefit greatly from such a new training capability.[64]

Standardize Simulators and Training Environments

The current simulators and synthetic training environments are not standardized across platforms. Different contractors build the simulators in which Air Force operators train—and build them differently.[65] This results in different levels of fidelity between simulators for different platforms.[66] Also, the security classification level of the synthetic training environment varies from platform to platform, with, for instance, fourth-generation aircraft having different levels of classification from fifth-generation aircraft. Investments in standardizing synthetic training environments could translate into a unified set of readiness metrics that are comparable across synthetic training environments and that would enable decisionmakers to better understand the aggregated readiness of a combined force package.

Furthermore, it is important for operators to be able to rehearse complex missions in a common environment that has the right threat and security classification levels. While live training ranges provide operators a shared training environment that is the same for all trainees, with specific capabilities and threat emitters that are known to all involved, this is not the case with current simulator training, even when the same capability assessment is being considered.

---

[62] Discussion with Air Force official, May 17, 2021.

[63] Discussion with Air Force official, May 17, 2021.

[64] Discussion with Air Force official, June 11, 2021. AMC has four core missions: airlift, air refueling, air mobility support, and aeromedical evacuation.

[65] Discussion with Air Force official, May 19, 2021.

[66] We use this definition of *fidelity:* "the degree to which the simulator replicates reality" (Tony Vonthoff, "The Importance of Fidelity in Simulation and Training," webpage, Modern Military Training, August 22, 2017).

Hence, the ensuing readiness assessments are not uniform or comparable across synthetic environments. Under a standardized synthetic environment there would be a common baseline for training across all weapon systems so that fidelity would not be an issue across synthetic environments and systems would also work across multiple security classification levels.[67]

Modernize Information Technology Infrastructure to Support Connectivity

Investments in IT infrastructure (including in platforms and connectivity) would also help improve readiness metrics. According to one senior leader, at the time of the interview, a strategic choice had yet to be made to make such investments. Such a strategy is needed and needs to be aligned with the Air Force's Digital Engineering Strategy. Furthermore, upgrading legacy platforms to the parameters required for operations in the current LVC environment would be needed because legacy platforms were not designed to accommodate agile software development. Finally, the senior leader remarked that investments in fifth-generation wireless technology represent the most-immediate investments needed in IT infrastructure to improve future readiness.[68] The reference to fifth-generation technology acknowledges that participants in future training environments would experience fewer environmental inconsistencies and lag with the higher bandwidth that this technology would provide.

However, another senior leader acknowledged the challenges that the lack of connectivity equipment and the fiscal realities the Air Force is facing pose: "Fiscal reality means that we won't get the connectivity equipment we want in the next budget, but in the next six to nine years"; if the Air Force were to wait the six to nine years to get connectivity equipment, "[we] will be woefully behind."[69]

Given the realities of the U.S. government and DoD budget cycles, the Air Force needs to make deliberate decisions to begin making investments in suboptimal ways (that is, not fully developed from a network security perspective) to avoid falling behind its great power competitors while waiting for an optimal way (that is, a fully developed implementation model from a network security perspective) to invest in IT infrastructure. In the senior leader's perspective, "this is an MVP [minimum viable product] model," and the Air Force needs to move toward making investments in updating and upgrading IT infrastructure at a suboptimal level instead of waiting for an "optimal way," if the United States intends to remain competitive in the great power competition game.[70]

---

[67] Discussion with Air Force official, May 19, 2021.

[68] Discussion with Air Force official, May 14, 2021.

[69] Discussion with Air Force official, June 11, 2021.

[70] Discussion with Air Force official, May 14, 2021.

Investments in More Simulators

Additional simulators will be needed to support more training being completed in simulators and the expansion of simulator training to include more participants across all services. One senior leader with whom we spoke emphasized the "unmet need for sims"—that not enough simulators are available for training, especially devices with high fidelity and concurrency.[71] For example, B-52, B-2, and B-1 simulators have shortfalls in fidelity, and simulators in Air Force Special Operations Command struggle with concurrency because of the high rate at which the command's aircraft are modified.[72] In the course of our discussion, the same senior leader said that in "an ideal world, we'd be getting more [readiness data] from sims than from aircraft. The outcomes that we want to predict are not available in live-fly."[73] This underscores the potential contribution simulators can make to assessing readiness.

Another senior leader mentioned the need for expanding the number of available simulators and expressed a preference for investments directed toward simulators rather than "engines" (aircraft). As engines are currently perceived to be a top priority, securing funding for them is easier than for simulators. The senior leader further explained that investments in simulators and simulator training are fundamental, and without fundamental simulator training for their operators, "it doesn't matter if you have engines or not, because you will fail at the mission."[74]

Investments in more simulators would primarily help address readiness assessment Gap 3, allowing more training in a wider variety of scenarios.

Investments in New Synthetic Threat Environments

The need for investments in additional synthetic threat environments for training came up in several of our discussions with Air Force senior leaders across MAJCOMs. For AMC operators, DMO training—which mostly relies on virtual and constructive components—currently represents the only way to receive the training the operators need to be able to respond to the threats outlined in the NDS.[75] While ACC operators have access to ranges for live training, the limited availability of ranges and of range time means that range access for the training of ACC

---

[71] Discussion with Air Force official, May 14, 2021. The Air Force Operational Training Infrastructure Flight Plan defines *simulator concurrency* as "the degree to which the synthetic environment correctly represents the real world" (USAF, "Air Force Operational Training Infrastructure 2035 Flight Plan," September 5, 2017. Also see "Simulator Concurrency: Why Military Operators Know It's Important to Winning the Fight," webpage, Modern Military Training, March 30, 2020.

[72] Discussions with Air Force officials, February 11, 2021, and February 16, 2021.

[73] Discussion with Air Force official, May 14, 2021.

[74] Discussion with Air Force official, June 11, 2021.

[75] *DMO* is defined as the "networking of warfighter training that uses the integration of virtual and constructive entities, systems, and environments via secure wide-area network to acquire and sustain mission essential competencies required for operational readiness" (AFI 16-1007, *Management of Air Force Operational Training Systems*, October 1, 2019, p. 19).

operators takes precedence over access for the training of AMC operators, who only "fly against threats in a synthetic environment."[76] At AFGSC, we were told that operators have a high requirement for simulator training "because they need to fight in any AOR [area of responsibility] on any day"[77]—a wide range of threat environments.

Data on how aircrew perform in different environments are valuable and can provide decisionmakers with insights on how well participants are performing in certain areas that have implications for readiness.[78] In this context, investments are needed to support an increase in synthetic threat emitters and threat environments to satisfy training requirements for operators across several MAJCOMS. Such investments would help address some of the concerns outlined in the discussion of readiness assessment Gap 3 mentioned in Chapter 3.

Investments in Aggregated Force Readiness Measurement

Investments in IT infrastructure to automate the measurement of aggregate force readiness data would facilitate reporting of aggregate readiness and address the concerns identified in Chapter 3 for Gap 2.[79] One Air Force senior leader mentioned that, at least at the squadron level, reporting of aggregated readiness metrics would be more useful than reporting of individual readiness. Furthermore, if readiness is automatically measured during training exercises, especially those that involve multiple units and platforms, data could be collected on factors that come into play only when forces are operating as an integrated team. This would also produce readiness measurements of forces as they would deploy rather than aggregating the readiness of individual pieces, improving the readiness information pipeline to senior leaders.

As mentioned in Chapter 3, squadrons associated with the Air National Guard and with the Air Reserve component report readiness through different commands, even though they are operating the same weapon systems.[80] Currently, the different reporting channels make it difficult to arrive at a reliable metric for aggregated readiness. Aggregated readiness reporting at air expeditionary wings is not possible in the DRRS-S system. According to one senior leader "DRRS[-S] doesn't have a good way to measure readiness at that level."[81]

Hence, improvements in aggregated force readiness measurement, possibly even including automation, would help address the gap in aggregated readiness reporting at higher echelons than the squadron (i.e., groups, wings), about which senior leaders care most. At the same time, these

---

[76] Discussion with Air Force official, June 11, 2021.

[77] Discussion with Air Force official, May 14, 2021.

[78] Discussion with Air Force official, May 19, 2021.

[79] The ability to measure aggregated force readiness is predicated on having events, either large-scale exercises or DMO training, where the measurement takes place. Thus, this investment is inherently tied into other training investments.

[80] Discussion with Air Force official, May 19, 2021.

[81] Discussion with Air Force official, May 19, 2021.

new measurement capabilities could capture metrics related to force integration, which would also address Gap 1, as presented in the Chapter 3.

Investments in Adaptive, Proficiency-Driven Training

One senior leader mentioned that investments in LVC assets that promote and support a training environment that is focused on and adapts to each individual trainee would help improve readiness metrics. The senior leader mentioned the importance of LVC assets that could identify and learn individual trainee patterns and adapt to the learning style and needs of each trainee. For instance, when a trainee has quickly acquired the skills associated with flying in daytime but is still struggling to achieve proficiency in flying at night, LVC assets would identify how quickly the trainee becomes proficient in each area. Using individual proficiency metrics, an adaptive LVC training system would then suggest a training schedule for the trainee that reduces the time the individual spends on day flights and increases the number of night flights that the individual needs to perform to attain proficiency.[82]

Additionally, training requirements are currently structured as a set number of sorties to gain proficiency, which may not ensure proficiency for all pilots. One pilot may be able to demonstrate flying proficiency in ten sorties, while another may need only three sorties to become proficient. A move toward a true proficiency-based assessment of readiness would contribute toward more-efficient use of training resources and make readiness assessment more objective.[83]

Investments in adaptive LVC assets that advance an individual-focused training environment would not only contribute to improving training in general but would also have the potential to facilitate a move toward a proficiency-based assessment of readiness for teams and teams of teams. For example, the same capabilities needed to assess the execution of tactics, techniques, and procedures (TTPs) can be used to assess how that execution affects team-level success. In addition to adapting the assessment capabilities to team-level assessment, improving individual training quality can also improve overall readiness understanding by helping identify areas across the force where individual skills are lacking.

Increased efficiency in the use of training infrastructure may also free training resources for integrated training. Current readiness metrics show whether a unit can execute its missions considering certain conditions and environments it might face. These readiness metrics do not address the proficiency with which a task is executed for a mission. Improvements in individual-level proficiency can translate into improved capability readiness at the aggregated UTC and force package levels and would benefit Gaps 1 and 2, especially if proficiency-based assessment expands beyond the individual to also assessing teams. However, we note that the development

---

[82] Discussion with Air Force official, May 14, 2021.

[83] Discussion with Air Force official, April 27, 2021.

of proficiency-based training assessment is not a mandatory precursor of addressing Gaps 1 and 2.

Implication

Four areas of investments in LVC assets that the Air Force senior leaders we interviewed deemed important could address the three gaps in readiness assessment previously presented to varying degrees. The first two, DMO training and synthetic threat environments, take the OTTI improvement pathway that improves the ability to rehearse operational scenarios and provides commanders with better information to assess readiness, although assessments may remain subjective. This follows the logic that practicing more scenarios better will increase readiness. The second two investments, measuring aggregated force readiness and adaptive, proficiency-driven training, relate to the second OTTI improvement pathway, which increases opportunities for data collection and makes readiness metrics more objective. The importance of the measurement aspect is that it is not enough to theoretically improve readiness, strategic decisionmakers needs to *know* how ready forces are through rigorous assessment.

5. The Common Synthetic Training Environment and Other Technologies on the Horizon

Numerous technologies in development are relevant for addressing gaps in readiness assessment and align with the LVC investments the four senior leaders recommended in Chapter 4. In this chapter, we examine several select areas, beginning with an ongoing Air Force effort called CSTE and discussing the technological challenges it may face and the implications for readiness assessment. Then, we review additional technological developments, identified during our literature review, that are related to team training and performance assessment.

The Common Synthetic Training Environment

The Air Force Agency for Modeling and Simulation is developing concepts for environments to advance tactical air training capabilities for fourth-, fifth-, and sixth-generation platforms.[84] The proposed environments are intended to close existing training capability gaps and augment the Air Force's capability to train effectively for future operating environments by introducing baseline changes to training infrastructure. Key goals include broader support for integrated fourth-, fifth-, and sixth-generation training; mitigating interoperability issues stemming from multiple levels of security; accurate representation of the electromagnetic operating environment (EMOE); improved capacity to represent emerging offensive capabilities and range of operational settings; direct support for joint tactical air training across all joint training tiers; and increased accessibility of tactical training systems. These goals represent a mixture of solutions for training gaps and technological limitations of current training capabilities.

JTSTE emerged directly from the Joint TACAIR Synthetic Training analysis of alternatives as a hybrid solution to address training gaps. As initially conceived, JTSTE comprises three major components, each making distinct contributions: the JITC, JSE Linked, and the Joint Data-Centric Synthetic Environment (JDC-SE). CSTE now replaces JDC-SE. However, the functionality and roles of CSTE and JDC-SE are otherwise identical:

- **JITC:** a small-unit training environment leveraging existing capabilities to enhance training
- **JSE Linked:** a platform-agnostic small-unit or larger-force advanced training environment (extends original JSE function and purpose)
- **CSTE:** a distributed, platform-agnostic large-force training environment

---

[84] We specify the three generations of platforms because other efforts do not include fourth-generation platforms and instead focus primarily on future platforms.

Because of cost and schedule constraints, CSTE is the only component being pursued as of this writing and represents the most comprehensive, albeit longer-term, solution proposed. The OTTI Synthetic Test and Training Capability defines several lines of effort to satisfy OTTI priorities more comprehensively (beyond the scope of JTSTE) and includes CSTE as one of the lines of effort.[85]

How CSTE Could Improve Readiness Assessment

As envisioned, CSTE will directly contribute OTTI-LVC improvements in three broad categories: training infrastructure, simulation capabilities, and data capture and storage.[86] CSTE is intended to provide a platform-agnostic design that centralizes the computing capabilities required for incorporating synthetic models in the training environment and better enables distributed training capabilities and readiness for joint and multidomain operations.[87] This approach intentionally shifts the focus of training capabilities away from system-specific simulators to a modular, open architecture that directly supports integrated training across air platforms.

CSTE's design goals require advances in training infrastructure for large-force, distributed training exercises to address interoperability issues arising from the secure exchange of information over networks and across classification levels (multiple levels of security). Network authorizations establish the methods and mutual approval between network operators for the transmission of data according to active cybersecurity policies. Therefore, CSTE is intended to establish and maintain network authorizations for recurring use, and simulation data will be centrally managed at the appropriate classifications for exchange between platforms engaged in training. Similarly, large-force training exercises require the timely and coherent exchange of information among training participants to convey realism during training. CSTE is intended to perform computations centrally and distribute the effects to simulators, where the resulting synthetic representation will be presented to trainees for real-time response.

These capabilities will rely strongly on data repositories residing within CSTE itself (e.g., centralized, standard models of Blue and Red air capabilities and training-relevant simulation data for environmental effects),[88] which will further promote concurrency of simulated platforms

---

[85] Nick Yates, "OTTI Update: Synthetic Test and Training Capability," NTSA Simulation & Training Community Forum 2021, August 11, 2021.

[86] Broadly speaking, OTTI includes technology for competency models, infrastructure, simulation capabilities, data capture and storage, and data analysis. Technological limitations in each of these categories affect the advancement of OTTI technologies in support of improved training and readiness assessment. See Toukan et al., 2022.

[87] By design, CSTE addresses computational challenges that represent limitations in the technical architectures of current training capabilities and their capacity to synthetically represent a variety of emerging Blue capabilities and Red threats, as assessed by the Joint TACAIR Synthetic Training analysis of alternatives.

[88] CSTE design requirements are intended to natively support a variety of models for existing and future air platforms such that it is computationally feasible to scale and integrate multiple, distinct platforms for synthetic training while preserving requirements for multiple levels of security.

with their real-world equivalents by centrally managing updates to the platforms being trained. Data resulting from training events will be captured and available for analysis and use in future training.

Presently, CSTE identifies several broad objectives with respect to OTTI:

- providing a standardized synthetic training environment that also normalizes software tools for training scenario development, integration and interoperation between simulated systems, synthetic training data preparation, and data management tools (including individual performance data)
- enabling a scalable distributed training environment that matches real-world capabilities and that is available at a pace matching combat readiness needs
- developing an enterprise-based training delivery capability that includes planning, briefing, and debriefing to enable flexible training related modeling and simulation systems and interorganizational and cross-community training collaboration
- providing integrated, multitier distributed training for current and future training platforms that natively addresses multilevel security and uses centralized computing to improve the secure exchange of relevant information between simulated platforms
- addressing operational training gaps, especially for next-generation and combat air force platforms, that stem from limitations of existing training infrastructure (e.g., range size constraints, representation of adversarial threats and capabilities, interoperation)
- expanding training capabilities to rapidly and realistically incorporate representations of advanced threats
- portraying a realistic synthetic training environment that is operationally relevant
- improving training proficiency and readiness for joint operations using state-of-the-art technological architectures.

When combined, these technological objectives are intended to contribute advanced training capabilities to OTTI that address the most pressing training objectives (e.g., fifth-generation training within an EMOE) within an extensible, government-owned environment. By design, CSTE is intended to provide the means to holistically address critical technological gaps presently limiting the value of training environments (live or synthetic) that are available today and to augment the Air Force's capacity to expand training beyond what is currently possible within existing environments.

CSTE's Approach to Training Delivery

CSTE's design and planning will be multipronged and will occur in parallel across key aspects of design. The implementation (spanning enabling infrastructure and the training environment on which it is built) will initially establish a host cloud to centralize training environment services. Distributed training will be enabled across multiple sites and be operable either for distributed training exercises or for training conducted in a disconnected, local-only mode. High-performance networking with authorizations will also be a CSTE design requirement to support the training experience in both speed and accuracy of feedback provided.[89]

CSTE will enhance existing training capabilities to support a broader mixture of platforms for integrated force training. However, existing simulators and simulation systems will require modification for use with CSTE. Integrated force training is expected to support interoperability with live components.

CSTE will support existing training performance analysis systems and methods through exportable performance data with improved granularity of training-relevant data and artifacts. CSTE will contribute to team-based performance measures only to extent that individual performance metrics inform team-based behaviors contributing to readiness. Although not an explicit CSTE requirement, team training that includes human-autonomous systems is expected to be accommodated by CSTE's design through extensible functionality. Similarly, CSTE will not include requirements for new development of training performance analysis systems or performance measurement devices but will support integration to performance measurement and readiness reporting systems.

CSTE is expected to provide authentic representations of space, cyber, electromagnetic, and weather effects with the goal of enabling realistic and operationally relevant training against peer and near-peer threats.

CSTE integration with existing legacy training systems via Simulator Common Architecture Requirements and Standards (SCARS) and newer training environments, such as the JSE, will likely require modifications to the legacy systems.[90] However, the SCARS effort is part of the Common Simulator Architecture line of effort (a part of the Synthetic Test and Training Capability), and it is a strategic goal for CSTE and SCARS architectures to be compatible.[91] The integration of joint or international partners systems for multidomain training might require additional, simulation-specific systems to support their inclusion in large-force training exercises. This depends on the extent to which partner training environments rely on compatible architectures for the execution of large-force training.

---

[89] Authorizations include the authority to operate pursuant to the DoD cybersecurity risk management policy and cybersecurity paradigms, such as implemented for zero-trust architectures (DoD Instruction 500.01, *Cybersecurity*, March 14, 2014, incorporating change 1, October 7, 2019; DoD Instruction 8510.01, *Risk Management Framework for DoD Systems*, July 19, 2022; Defense Information Systems Agency and National Security Agency Zero Trust Engineering Team, *Department of Defense (DoD) Zero Trust Reference Architecture*, July 2022).

[90] SCARS is a sustainment initiative for enterprise management of existing legacy training systems based on a common open systems architecture; see John Kurian, "Simulators Division (AFLCMC/WNS) Update," briefing slides, June 16, 2021. See also Frank Wolfe, "U.S. Air Force Joint Simulation Environment for F-35A, F-22, Other Platforms to Build on F-35 JSE," *Defense Daily*, July 23, 2021.

[91] Kurian, 2021.

Technological Challenges and Implications for Readiness

The purpose of CSTE is to advance multiple areas of OTTI capability simultaneously by addressing concrete technical challenges using state-of-the-art methods using a comprehensive approach. Correspondingly, a range of technical challenges with implications for readiness assessments are being addressed.

Examples of the technical challenges and their potential implications are associated with the three readiness assessment gaps in Table 5.1. Similar to the readiness assessment gaps, the example technical challenges are also associated with distinct areas of OTTI capability, such as scalability, interoperability, authenticity, realism, concurrency, training delivery, and performance assessment.[92]

Many of the technical challenges have implications for more than one readiness assessment gap:

- Addressing the technical interoperability required to enable accurate force configurations across simulated platforms and enabling greater authenticity in the representation of events and effects presented to warfighters by the synthetic training environment combine to affect all three readiness assessment gaps.
- Even with sufficient technological advancements enabling other OTTI capabilities, the closing of readiness assessment gaps will remain limited by the availability of new performance measurement tools, specialized data-collection systems, and methods to represent critical concepts for the operational force.
- Emerging command, control, and communications–related operating concepts also need training delivery capabilities, with associated performance assessment concepts to systematically assess force readiness, including factors of integration and aggregated force readiness.
- Addressing each technical challenge contributes to closing a readiness assessment gap but not necessarily independently. For example, addressing limits on the number of simultaneous training participants (scalability) and the accessibility of suitably low-latency, high-capacity, and long-distance networking (training delivery and interoperability) would simultaneously enable larger-scale force training exercises and broaden the training accessible to home stations participants.
- OTTI capabilities enabled by addressing the technical challenges presented by accurate representation of an EMOE, multiple levels of security for the exchange of information, streamlined development tools for the preparation and execution of training exercises, and sensor-based data collection to inform performance measurements are also key enablers for CSTE that would close readiness assessment gaps.

The next section reviews literature concerning additional technological developments that are potentially relevant for closing readiness assessment gaps and CSTE's design.

---

[92] The glossary provides definitions for these terms.

Table 5.1. CSTE Technical Challenges and Readiness Assessment Implications

Affected Areas of OTTI Capability	Example Technical Challenge	Example Implication for Readiness Assessment	Readiness Assessment Gap(s)
Scalability	Limits on the number of simultaneously supportable participants	Gap in information on higher-level readiness Training throughput	Integration Aggregation
Authenticity	Standardized, accurate representation and distribution of effects in an EMOE	Inaccurate portrayal of and assessment within foreseeable threat environment Inconsistent effects across simulated platforms	Integration Aggregation Scenario
Authenticity and realism	Appropriately sequenced events, synchronized environmental effects, and availability of multiresolution, multidomain environmental data	Partial insight about the operational impact of capabilities in large-force, joint operations	Integration Aggregation Scenario
Concurrency Realism	Highly accurate digital representation of weapon platforms in function and form	Training for TTPs does not transfer to operational settings	Scenario
Interoperability Authenticity	Integration with existing training federations and non-U.S. partner training environments	Limited insight into readiness for joint or combined operations resulting from large-force training exercises	Integration Aggregation Scenario
Interoperability	Multiple-level security constraints	No readiness feedback on how fifth-generation integrates with other platforms	Integration Aggregation
Interoperability Training delivery	Integration with live assets	Limited or no ability to extend the reach of live training exercises to include home station participants using remote synthetic training environments	Integration Scenario
Interoperability Training delivery	Low-latency, high-capacity networking over long distances	Training may be limited to home stations in specific geographic regions	Integration Aggregation Scenario
Training delivery	Flexible and streamlined development tools for the preparation and execution of large-force training exercises across joint and combined participants	Technical issues limit the efficient execution of mission sets during training Inconsistent scenario implementations across training platforms	Integration Scenario

36

Affected Areas of OTTI Capability	Example Technical Challenge	Example Implication for Readiness Assessment	Readiness Assessment Gap(s)
Performance assessment	Performance analysis tools enabled for real-time assessments, including team-based and cognitive load evaluations	Readiness assessments will be limited by performance analysis tools associated with past training capabilities or to platform-specific innovations with limited insight	Integration Aggregation
Performance assessment	Sensor-based performance data collection for assessment and feedback	Limited ability to maximize team and individual proficiency Status quo on measuring performance	Integration Aggregation
Training delivery Performance assessment	Methods to represent emerging operational command, control, and communications systems synthetically	Emerging Blue force tactics and operational command-and-control concepts will have limited or no representation in larger-scale force exercises	Integration Aggregation

NOTE: Definitions for the listed areas of OTTI capability are given in the glossary. The readiness assessment gaps refer to those discussed in Chapter 3 as follows: Gap 1 = Integration; Gap 2 = Aggregation; Gap 3 = Scenario.

Other Technological Developments

We came across other technological developments in our literature review that are relevant to addressing readiness assessment gaps (see Chapter 3). These developments also support recommendations from four senior leader interviews for investments in DMO training, measurement of aggregated readiness, and adaptive training (see Chapter 4). These developments have potential for being incorporated in CSTE but may also be a part of other new training infrastructure investments. The reviews of technological developments in this section are from the perspective of team training and performance assessment. This theme was chosen because it touches on the gaps and investment recommendations from previous chapters and is also a relatively underdeveloped area for DoD. That said, the Air Force does already train aircrews together in multiple simulator events. In our discussion of improving and building team training and assessment, we refer both to expanding capabilities to support aircrew training activities and scaling up the team to the force package or joint task force size.

Principles for Team Training and Performance Assessment

As the Air Force invests in technologies that create new opportunities for teams (or teams of teams) to train together in synthetic environments, efforts to capitalize on the digital byproducts should draw on key principles for assessing the performance of teams. A recent meta-analysis, rooted in nearly 30 years of research on team training and performance evaluation, proposed five principles for team-focused simulation training.[93]

First, any effort to improve team dynamics through training must have **clear objectives and performance standards (OPS)**. These should be rooted in the specific knowledge, skills, and abilities an organization would like to improve. It is also important to **use team-oriented (not individual-oriented) simulation-based trainers**. While this may seem intuitive, aggregated individual training results are sometimes used to assess team performance, as pointed out in our discussions with four Air Force senior leaders (see Chapter 4). The point of this principle (and a subject examined in other research) is that training systems designed specifically around team-level metrics are better indicators of team performance than those based on the combination of individual-level results.[94]

---

[93] Margaret Thomson Crichton, "From Cockpit to Operating Theatre to Drilling Rig Floor: Five Principles for Improving Safety Using Simulator-Based Exercises to Enhance Team Cognition," *Cognition Technology & Work*, Vol. 19, No. 1, 2017.

[94] Sean L. Normand and Joan H. Johnston, "A Qualitative Study on Behavioral Markers of Team Cohesion and Collective Efficacy to Inform the Army's Synthetic Training Environment," paper presented at 2020 Virtual Interservice, Industry, Training, Simulation, and Education Conference (vIITSEC), National Training and Simulation Association, 2020; Steve W. J. Kozlowski and Georgia T. Chao, "Unpacking Team Process Dynamics and Emergent Phenomena: Challenges, Conceptual Advances, and Innovative Methods," *American Psychologist*, Vol. 74, No. 4, May–June 2018.

The third principle is that assessments of team performance should **use behavioral markers for evaluation**. That is, metrics should be designed in such a way that a training observer can evaluate performance by examining individual and team behaviors that participants display during training. For training to be effective, the effort must also **provide prompt performance feedback to teams.** Finally, **simulation-based training should be conducted regularly**. This subsection discusses procedural and technological developments across these five principles.

Objectives and Performance Standards

Effective team training OPS are clearly defined and anchored in a set of knowledge, skills, and abilities that the organization would like to target for improvement (generally identified through a competency or task analysis).[95] Developing OPS can potentially be a lengthy process and require many resources, including the guidance of SMEs with specific knowledge about developing effective OPS. It also requires iterative testing to ensure that the OPS reasonably capture the desired improvements.

To simplify this process, tools have been developed that provide systematic methods for developing team training OPS specific to an organization's goals. For example, the Scenario-Based Performance Observation Tool for Learning in Team Environments has been used to develop and demonstrate behavior-linked OPS for four-person F-16 pilot teams training for air-to-air combat in a simulated environment.[96] Other teamwork behavior-linked methodologies, such as Targeted Acceptable Responses to Generated Events or Tasks and the U.S. Navy's Teamwork Observation Measure, were developed specifically for use in distributed mission training.[97] The DAF could leverage these or similar tools to create and adapt measures of aggregated readiness, as recommended by Air Force senior leaders.

Another recommendation from the senior leaders we interviewed was the adoption of adaptive training for individuals. Adaptive training (also called intelligent tutoring) uses artificial intelligence to automatically adjust OPS according to trainee behavior and performance learning goals. The technology for adaptive training is well-developed and continues to grow. Within the simulation-based training research community, interest has been growing in adaptive training, and many organizations (including branches of the U.S. military) are continuing efforts to develop these capabilities. Some tools, like Dignitas Technologies' Flexible and Live Adaptive Training Tool (which is integrated with the Army's Synthetic Training Environment) even allow human operators to make live changes to simulation-based training during observation of an

---

[95] Crichton, 2017.

[96] Jean MacMillan, Eileen B. Entin, and Rebecca Morley, "Measuring Team Performance in Complex and Dynamic Military Environments: The SPOTLITE Method," *Military Psychology*, Vol. 25, No. 3, 2013.

[97] MacMillan, Entin, and Morley, 2013; Ebb Smith, Jonathan Borgvall, and Patrick Lif, *Team and Collective Performance Measurement*, DSTL, Policy and Capability Studies, June 30, 2007.

event.[98] Adaptive training systems have been created for individual training in computer science, languages, math, physics, cloud computing, and many other fields.[99] Adaptive training for collective and team training is a newer and less established field; however, significant research is also being done on this front, notably with the U.S. Army Research Lab's Generalized Intelligent Framework for Tutoring (GIFT).[100]

Team-Oriented Simulation-Based Trainers

Some measures of teamwork rely simply on the aggregation of individual training results; however, research has shown that team-specific simulation-based trainers are better indicators of team performance and teamwork.[101] This echoes the concerns and recommendations that we heard from our interviews with Air Force senior leaders. Like the development of team training OPS, however, creating team-based trainers can be a lengthy and expensive process. Accordingly, methods and tools have also been developed to make this process simpler and more accessible. GIFT is one notable example and was originally designed to simplify the creation of individual adaptive training systems.[102] Several recent efforts have explored using GIFT to create team performance trainers. Team-based studies from the 2021 GIFT Users Symposium include how to define user roles and relationships for team training, and one research group has demonstrated a preliminary design for a distributed team trainer for senior leader decisionmakers.[103]

Teamwork Behavioral Markers

Air Force senior leaders we interviewed also showed concern that most (if not all) current aggregated capability readiness metrics are based on subjective evaluation. To reduce subjectivity, readiness metrics related to integration and larger, aggregated teams should be

---

[98] Steven Harrison and Elyse Burmester, "FLATT: A new Real-Time Assessment Engine Powered by GIFT," in Benjamin S. Goldberg, ed., *Proceedings of the Ninth Annual GIFT Users Symposium*, May 2021; Dignitas Technologies, "Dignitas Awarded Flexible and Live Adaptive Training Tools (FLATT) Phase II SBIR," webpage, February 8, 2021.

[99] Alaa N. Akkila, Abdelbaset Almasri, Adel Ahmed, Naser Al-Masri, Yousef Abu Sultan, Ahmed Y. Mahmoud, Ihab Zaquot, and Samy S. Abu-Naser, "Survey of Intelligent Tutoring Systems up to the end of 2017," *International Journal of Academic Information Systems Research*, Vol. 3, No. 4, April 2019.

[100] Robert A. Sottilare, C. Shawn Burke, Eduardo Salas, Anne M. Sinatra, Joan H. Johnston, and Stephen B. Gilbert, "Designing Adaptive Instruction for Teams: a Meta-Analysis," *International Journal of Artificial Intelligence in Education*, Vol. 28, June 2017; Benjamin Goldberg, ed. *Proceedings of the Ninth Annual GIFT Users Symposium*, May 2021.

[101] Normand and Johnston, 2020, p. 10; Kozlowski and Chao, 2018.

[102] GIFT Tutoring, "Generalized Intelligence Framework for Tutoring (GIFT)," homepage, undated.

[103] J. T. Folsom-Kovarik, Joel Sieh, and Anne M. Sinatra, "Reasoning About Team Roles and Responsibilities for Team Assessment," in Benjamin S. Goldberg, ed., *Proceedings of the Ninth Annual GIFT Users Symposium*, May 2021; Randy Jensen, Randy, Bart Presnell, Jeanine DeFalco, and Gregory Goodwin, "Designing a Distributed Trainer Using GIFT for Team Tutoring in Command Level Decision Making and Coordination," *Proceedings of the Ninth Annual GIFT Users Symposium*, May 2021.

based on teamwork behavioral markers; that is, measurable, observable behaviors, such as the physical movement of team members or the type of language the team uses. Markers should be validated to ensure they capture and represent the desired elements of teamwork.[104]

A 2020 study investigated a set of behavioral performance markers in air combat, including shared situation awareness, mental workload, and normative performance of TTPs, finding that these are better indicators of effective teamwork than simply measuring performance output.[105] Rather than a simple indicator of whether the operational task was completed successfully, these additional metrics provide analysts insight into whether the team communicated effectively (shared situational awareness), how taxing the operation was on individuals (mental workload), and whether proper TTPs were followed, regardless of the overall success of the operational task. These metrics do provide a fuller picture of training and team *outcomes*, but, apart from shared situational awareness, they do not directly contribute to an understanding of team dynamics.

In recent literature, team cohesion, team efficacy, and team communication are cited as three major metrics of team dynamics, and many studies use behavioral markers that align with these metrics (even if they are not expressly categorized in this way), including in military contexts.[106] Table 5.2 maps team behavioral markers from notable recent studies across these metrics. In some studies, these three foundational metrics are used to inform complex team dynamics, such as team connectivity (how actions and individuals act as part of a whole rather than in isolation), team functional resilience (how well teams remain effective after losing a piece of functionality), and team uncertainty (the cognitive state of a team based on the cognitive states of its constituent individuals).[107]

---

[104] Crichton, 2017.

[105] Heikki Mansikka, K. Virtanen, D. Harris, and M. Jalava, "Measurement of Team Performance in Air Combat—Have We Been Underperforming?" *Theoretical Issues in Ergonomics Science*, Vol. 22, No. 3, 2021.

[106] Shannon L. Marlow, Christina N. Lacerenza, Jensine Paoletti, C. Shawn Burke, and Eduardo Salas, "Does Team Communication Represent a One-Size-Fits-All Approach? A Meta-Analysis of Team Communication and Performance," *Organizational Behavior and Human Decision Processes*, Vol. 144, 2018; Susan H. McDaniel and Eduardo Salas, "The Science of Teamwork: Introduction to the Special Issue," *American Psychologist*, Vol. 73, No. 4, 2018.

[107] Zachari Swiecki, Morten Misfeldt, Xiangen Hu, and David Williamson Shaffer, "Visualizing Team Processes Using Epistemic Network Analysis: Affordances for Researchers, Educators, and Teams," in Anne M. Sinatra, Arthur C. Graesser, Xiangen Hu, Benjamin Goldberg, and Andrew J. Hampton, eds., *Design Recommendations for Intelligent Tutoring Systems*, Vol. 8: *Data Visualization*, U.S. Army Combat Capabilities Development Command, Soldier Center Simulation and Training Technology Center, 2020; Folsom-Kovarik et al., 2021; Ron Stevens, Ryan Mullins, Xiangen Hu, Diego Zapata-Rivera, and Trysha Galloway, "Visualizing the Momentary Neurodynamics of Team Uncertainty," in Anne M. Sinatra, Arthur C. Graesser, Xiangen Hu, Benjamin Goldberg, and Andrew J. Hampton, eds., *Design Recommendations for Intelligent Tutoring Systems*, Vol. 8: *Data Visualization*, U.S. Army Combat Capabilities Development Command, Soldier Center Simulation and Training Technology Center, 2020.

Table 5.2. Teamwork Metrics and Corresponding Behavioral Markers in Recent Literature

Metric	Behavioral Markers
Team cohesion	Taking initiative, contribution to problem solving, taking responsibility, seeking input, attitudes, affirmations[a]
	Initiative leadership, measured with "provide guidance" and "stay on mission"[b]
	Initiative or leadership (providing guidance, stating priorities)[c]
Team efficacy	Task confidence, soldier assistance, cooperation, conflict resolution, interpersonal tact[a]
	Supporting behavior, measured with "provide backup" and "request backup"[b]
	Supporting behavior[d, e]
	Providing cover for team members, avoiding fratricide[d, f]
	Supporting behavior (correcting errors, providing and requesting backup)[c]
Team communication	Shared situation awareness[g]
	Individual (e.g., body language, appropriate terminology, incorporating feedback, seeking input from team), team (e.g., team energy, use of team member names, team input at decision points)[h]
	Information exchange (anticipate information needs, provide situation updates), communication delivery (provide complete reports, use clear communications)[b]
	Information exchange, communication, advanced situation awareness[d, e]
	Awareness of team members and intent, information-sharing[d, f]
	Information exchange (using available sources, passing information before being asked, providing situation updates), communication delivery (using correct terms providing complete reports, using brief communications, using clear communications)[c]

[a] Normand and Johnston, 2020.
[b] Folsom-Kovarik et al., 2021.
[c] Joan H. Johnston, Anne M. Sinatra, and Zachari Swiecki, "Application of Team Training Principles to Visualizations for After-Action Reviews," in Anne M. Sinatra, Arthur C. Graesser, Xiangen Hu, Benjamin Goldberg, and Andrew J. Hampton, eds., *Design Recommendations for Intelligent Tutoring Systems*, Vol. 8: *Data Visualization*, U.S. Army Combat Capabilities Development Command, Soldier Center Simulation and Training Technology Center, 2020.
[d] Demonstrated automated measurement.
[e] Randall Spain, Wookhee Min, Jason Saville, Keith Brawner, Bradford Mott, and James Lester, "Automated Assessment of Teamwork Competencies Using Evidence-Centered Design-Based Natural Language Processing Approach," in Benjamin S. Goldberg, ed., *Proceedings of the Ninth Annual GIFT Users Symposium*, May 2021.
[f] Caleb Vatral, Naveeduddin Mohammed, Gautam Biswas, and Benjamin S. Goldberg, "Gift External Assessment Engine for Analyzing Individual and Team Performance for Dismounted Battle Drills," in Benjamin S. Goldberg, ed., *Proceedings of the Ninth Annual GIFT Users Symposium*, May 2021.
[g] Mansikka et al., 2020.
[h] Deanna L. Reising, Douglas E. Carr, Sally Gindling, Roxie Barnes, Derrick Garletts, and Zulfukar Ozdogan, "An Analysis of Interprofessional Communication and Teamwork Skill Acquisition in Simulation," *Journal of Interprofessional Education & Practice*, Vol. 8, No. 1, September 2017.

Provide Prompt Performance Feedback to Teams

Current Air Force training practices principally use trained observers or SMEs to evaluate behavioral markers in team training. SME evaluations have been found to be more accurate indicators of team performance than self-reported measures.[108] However, SME evaluation is

---

[108] Normand and Johnston, 2020.

limited by the availability of trained observers. Some Air Force senior leaders we interviewed recommended automating the collection and evaluation of team performance metrics. Following this recommendation would make training assessment more flexible and could allow more-efficient feedback to teams.

Although there are still many technological challenges, significant research is being done to advance automated team assessments, particularly around team communication. For example, Vocavio Technologies' vSIM software automatically assesses team dynamics by analyzing pitch, rhythm, and tone of voice between team members during simulated or live pilot training and operations.[109] Similarly, a 2021 study investigated using natural language processing technology to automatically characterize team behavioral markers in U.S. Army mission training.[110]

In a recent study, researchers demonstrated a tool that outputs automated performance assessments across a number of individual and team metrics.[111] This tool, the External Assessment Engine (EAE), was demonstrated in an Army dismounted battle drill, an exercise that requires complex skills and the ability to work effectively as part of a team toward a shared objective. EAE collects and synthesizes data from multiple sources: video of soldier movements, video of the battle drill scenario, and state data from the simulator related to soldier activity. Using these data, EAE automatically evaluates psychomotor skills, marksmanship, situational awareness, rapid decisionmaking, and team coordination. The tool also provides videos of training to instructors to simplify the after-action review process.

Conduct Regular Simulation-Based Training

The final principle for training team dynamics advocates regular simulation-based training. Air Force senior leaders we interviewed indicated that this may be an area of difficulty for the Air Force. This principle is tied to these senior leaders' recommendations for expanded and standardized simulated environments, more platforms, and developing IT infrastructure to connect platforms, all of which could make simulation-based training more available to individual and team training.[112] Investing in such technology as adaptive training to make current training more efficient could also potentially increase the rate of aggregated training. Current literature has not explored optimal rates of team training; this is an area that requires more research.[113] As with the other principles presented here, this is not an issue that will be automatically solved by fielding a capability like CSTE. It will take significant study and planning to optimize how the USAF will schedule different participants, and at what intervals, to

---

[109] Vocavio Technologies, "Vsim: Pilot Training," webpage, undated.

[110] Spain et al., 2021.

[111] Vatral et al., 2021.

[112] Here, we consider team training to be inclusive of training as a crew, a force package, or a joint task force.

[113] Crichton, 2017.

exercise and assess aggregated performance against relevant complex scenarios, which must then be aligned and incorporated into readiness reporting.

Implication

The development of CSTE has the potential for addressing the readiness assessment gaps discussed in Chapter 3. However, for those benefits to be realized, readiness assessment must be considered when making design decisions to overcome the technical challenges, such as those described in this chapter. If readiness assessment is not left as an afterthought, it can be an added return on investment for CSTE. The second part of this chapter, where we discussed other technological developments, highlights that solving the gaps in readiness assessment will require more than buying additional simulator technology and synthetic environments. Many aspects of our readiness assessment gaps relate to missions that are performed as a team, and attention must be paid to developing and adopting new concepts in team training and team performance assessment to best capitalize on OTTI investments that allow individuals to train together.

6. Recommendations and Conclusion

The challenges of determining strategies to meet new NDS priorities across DoD functions at all levels have produced a renewed interest in understanding how planning decisions will affect readiness. The current readiness system defines and measures two main types of readiness. Resource readiness assumes that unit design and structure are sufficient to generate capabilities and measures conformity with the design. Capability readiness, a supplementary concept introduced to capture outputs rather than inputs, asks unit commanders to subjectively assess the unit's ability to accomplish a list of METs.

While some information from the readiness reporting system is useful to inform certain decisions, the approaches to measuring capability readiness and the associated data are insufficient to answer cause-and-effect questions about strategic outcomes in potential future conflicts (as many senior leaders have noted).[114] But the Air Force is currently weighing investments in synthetic training environment technologies that have great potential for improving capability readiness information (as we describe in Chapters 4 and 5).

Our research suggests that whether or not planners can capitalize on this potential depends on many future decisions that determine the particular design and acquisition priorities of the synthetic environments, which are still in the very early stages of requirement definition. Yet synthetic environments alone (and the associated digital byproducts) will not create better information for decisionmakers unless the Air Force also addresses the structural flaws in the readiness reporting system (discussed in Chapter 3). The following recommendations will help improve capability readiness understanding in the short term while positioning the Air Force to capitalize on the longer-term possibilities that synthetic environments might enable.

Recommendations

Further differentiate capability readiness and align new dimensions with supporting inputs from appropriate functions at headquarters and MAJCOMs. For readiness to function as a useful objective for understanding cause-and-effect relationships, we suggest that readiness definitions should be specific, measurable, attainable, relevant, and time-bound (SMART).[115] Thus, if leaders desire to move from a narrow definition of readiness focused on the status of current forces to a broader definition, making the objective less specific, planners should add the specificity back in by creating differentiated subobjectives, as discussed in

---

[114] Brown and Berger, 2021; Austin, 2021, p. 12.

[115] Christopher Paul, Jessica Yeats, Colin P. Clarke, Miriam Matthews, and Lauren Skrabala, *Assessing and Evaluating Department of Defense Efforts to Inform, Influence, and Persuade: Handbook for Practitioners*, RAND Corporation, RR-809/2-OSD, 2015.

Chapter 2. Recent efforts by senior leaders in the DoD to enumerate the core elements of strategic readiness,[116] along with the Air Force introduction of the threat dimension into readiness reporting, point to the need to move in this direction of specifically defining the elements of the broader readiness definition.

However, readiness reporting still falls under the operations functional chain of command, where planners are most focused on the status of current forces. Broadening the definition beyond the status of current forces creates new demand for a broader set of inputs into the reporting process. While evaluating the adequacy of capabilities for wartime scenarios continues to have an operational component, accurate capability assessments under the broader definition potentially require inputs from intelligence (A2) channels on adversary capabilities, logistics (A4) channels for issues of sustaining capabilities in extended scenarios, plans and requirements (A5) channels for specificity on precise planning scenarios in view, and strategic plans and programs (A8) for the status of technologies in the acquisition pipeline and their potential impacts on readiness. In sum, the Air Force should define SMART elements of the broader definition and begin to align these elements with inputs from the appropriate functional authorities.

Consider a process mechanism to bring information into readiness reporting from more-appropriate sources when unit commanders lack information. The previous recommendation makes it clear that unit commanders, alone, are not in the best position to produce the cross-functional information necessary to generate informative capability assessments for today's strategic questions (see Chapter 3 gaps in readiness assessment). Even today, high-quality information exists on how a fully resourced UTC or force package will likely perform in a given operational scenario, but there is no mechanism for this information to influence capability readiness reporting.[117] We recommend that planners establish the pathways for the best available information from across different functions to inform readiness reporting. Then, any improvements to the state of capability knowledge enabled by new synthetic training opportunities can follow the same pathways to improve future readiness assessments.

Consider adding a field in DRRS-S to capture the quality of information used as inputs for subjective assessments. Adding such a field would be an immediate improvement to the data-collection approach. Specifically, our analysis of DRRS-S comments showed that commanders often include remarks qualifying their assessments based on limitations in the information available to them (sometimes because OTTI limitations do not allow them to observe realistic training on the particular MET), which relates to the scenario gap we introduced in Chapter 3. Collecting this information explicitly would provide feedback on the quality of information informing subjective assessments today but, more importantly, would position the

---

[116] Martin et al., 2021.

[117] The primary example of this is exercises where multiple UTCs practice operational scenarios. There will be a debrief and often an after-action report, both of which do not currently directly feed into official readiness reporting.

Air Force to measure the impact of new synthetic training capabilities on the quality of information flowing into the system (see discussion of Figure 4.1).

Create a working group focused on data and measurement to guide synthetic-environment design decisions. The first recommendation mentions the wide range of functional entities that touch on capability readiness. An equally wide range of entities stands to benefit from the general-purpose information that might be created and captured by future synthetic training environments. To ensure that new synthetic environments meet the diverse needs of these stakeholders, the Air Force should form a cross-functional working group to advise acquisition efforts on design issues pertaining to data and measurement. The goal of the working group would be to ensure design decisions (e.g., the design of CSTE, as discussed in Chapter 5) meet the needs of research and development users in addition to those involved in operational training.

Factor readiness assessment gaps into OTTI priorities. Finally, our research (see Chapters 4 and 5) shows that training technologies have significant implications for readiness assessment. Yet, plans and priorities for future OTTI capabilities might not realize the full benefit of the capabilities unless they factor in the impact of training technologies on the readiness assessment gaps we identified in Chapter 3. For example, the investments that enhance training for individual operators, such as greater degrees of realism, might come at the expense of broadening the set of platforms that can participate, which would be more targeted toward the readiness assessment gaps we identified. Planning documents, such as the OTTI flight plan, should consider the readiness benefits when setting priorities for OTTI development.

Conclusion

Producing informative readiness assessments is a long-standing defense challenge that is likely to intensify with the increasing technological complexity of the potential conflicts for which the NDS directs the Air Force to prepare. But synthetic training technologies, which are also improving, have the potential to support better assessments through improvements in the ability to simulate combat conditions. The best way forward is for the Air Force to create an institutional framework that brings the best available information into the readiness system, while doing its best to capitalize on new data when they become available.

Appendix. Senior Leader Discussion Protocol

General Questions

1. What is your working definition of readiness for the units under your purview?

    – What primary inputs do you factor into your working definition of readiness?

2. What elements of USAF or Joint/DOD readiness metrics are the most informative in assessing the readiness of units under your purview (IAW your working definition of readiness)?

3. Could you share your perspective on which aspects of readiness are most important and the different types of decisions you use them to make?

4. Based on your experience as a consumer of readiness metrics, which are the most useful metrics to inform those decisions?

    – Could you please elaborate on the reasons why you find them useful?

    – Along similar lines, which are the least useful metrics for decisions, and which are the reasons why you do not find them useful?

5. Do you encounter decisions that challenge you to weigh different aspects of readiness against each other (e.g., live-fly vs. upgrade simulators, or counterterrorism deployment taskings vs. OPLAN scenario requirements)? If yes, could you elaborate on those decisions and how you weigh trade-offs?

6. What are the main readiness-related issues that your command is experiencing?

7. Are there MAJCOM-specific unit reporting requirements you use in addition [or instead of] to USAF or Joint/DOD readiness metrics when assessing the readiness of units under your purview (according to your working definition of readiness)?

8. How and at what level (unit, MDS) do you typically consume readiness metrics and information?

9. When considering status of forces for a particular contingency or scenario, are there important aspects of readiness that are not captured by current reporting? If yes, what are those aspects and what are the reasons why they are not being captured by the current reporting process?

10. How do available LVC training assets or resources currently contribute to your understanding of readiness of the units under your purview?

11. What set of LVC training assets or resources have you found or consider to be most useful in the production of readiness metrics? Please elaborate why.

12. In your assessment, how could LVC assets contribute to improving the readiness inputs for your decisions?

    – Are there factors favoring the use of LVC or barriers to their use?

    – What are priority investments for improving use of LVC capabilities for readiness metrics?

13. What are some of your top recommendations regarding future improvements to readiness metrics in general, and the use of LVC to enhance readiness reporting specifically?
14. In closing, what have we not covered that we need to know?

Abbreviations

ACC	Air Combat Command
AFGSC	Air Force Global Strike Command
AFI	Air Force Instruction
AI	artificial intelligence
AMC	Air Mobility Command
ART	Air and Space Expeditionary Forces UTC Reporting Tool
CSTE	Common Synthetic Training Environment
DAF	Department of the Air Force
DMO	distributed mission operations
DoD	U.S. Department of Defense
DRRS-S	Defense Readiness Reporting System–Strategic
EAE	External Assessment Engine
EMOE	electromagnetic operating environment
GIFT	Generalized Intelligence Framework for Tutoring
IT	information technology
JDC-SE	Joint Data-Centric Synthetic Environment
JITC	Joint Integrated Training Center
JSE	Joint Simulation Environment
JTSTE	Joint TACAIR Synthetic Training Environment
LVC	live, virtual, constructive
MAJCOM	major command
MET	Mission Essential Task
NDS	National Defense Strategy
OPLAN	operational plan
OPS	objectives and performance standards
OTTI	operational test and training infrastructure
SCARS	Architecture Requirements and Standards
SMART	specific, measurable, attainable, relevant, and time-bound
SME	subject-matter expert
SORTS	Status of Resources Training System
TACAIR	tactical air
TTP	tactics, techniques, and procedures
USAF	U.S. Air Force
UTC	Unit Type Code

Glossary of OTTI-Related Terminology

authenticity | The preservation of contextual validity and the essential traits of a simulated scenario or group of entities.

authenticity and realism | Closely related terms expressing similar concepts for the quality of a simulated aspect of training within a synthetic environment.

concurrency | "The condition where the configuration and operation of the operational training system matches the configuration and functionality of the reference weapon system(s), to the extent necessary to provide required training. For training devices, this condition includes the operational training systems' operational [flight] program, mission software, weapons, hardware, and third-party systems that sufficiently and accurately reflects the current configuration of the weapon system(s) functionality."[118]

interoperability | 1. "The ability to act together coherently, effectively, and efficiently to achieve tactical, operational, and strategic objectives."[119]
2. "The condition achieved among communications-electronics systems or items of communications electronics equipment when information or services can be exchanged directly and satisfactorily between them and/or their users."[120]

performance assessment | An evaluation of training achievement based on standard training conditions, standard training methods, and accepted systems of measurement.

realism | The extent to which a simulated representation of an object or system of objects depicts or behaves in the same manner that the object(s) being represented would in a corresponding real-world setting.

scalability | "The ability of a distributed simulation to maintain time and spatial consistency as the number of entities and accompanying interactions increase."[121]

---

[118] AFI 16-1007, 2019, p. 19. Brackets in original.

[119] JP 6-0, *Joint Communications System*, June 10, 2015, Joint Chiefs of Staff, incorporating change 1, October 4, 2019, p. GL-5.

[120] JP 6-0, 2019, p. GL-5.

[121] DoD, *Department of Defense Modeling and Simulation (M&S) Glossary*, Modeling and Simulation Coordination Office, October 1, 2011, p. 143.

training delivery   The methods and means by which the training system is implemented and
                                         training is performed.

training system   "A systematically developed curriculum including, but not necessarily
                                         limited to, courseware, classroom aids, training simulators and devices,
                                         operational equipment, embedded training capability, and personnel to
                                         operate, maintain, or employ a system. The Training System includes all
                                         necessary elements of logistic support."[122]

---

[122] Air Education and Training Command Instruction 36-2621, *Flying Training Course Publications Development*, November 3, 2020, corrective action, October 15, 2020, p. 18.

References

Air Education and Training Command Instruction 36-2621, *Flying Training Course Publications Development*, November 3, 2020, corrective action, October 15, 2020.

Air Force Instruction 10-201, *Force Readiness Reporting*, December 22, 2020. As of August 23, 2021:
https://static.e-publishing.af.mil/production/1/af_a3/publication/afi10-201/afi10-201.pdf

Air Force Instruction 10-244, *Reporting Status of Air and Space Expeditionary Forces*, U.S. Air Force, June 15, 2012.

Air Force Instruction 10-244, *Reporting Status of Air and Space Expeditionary Forces*, supplement, U.S. Air Forces in Europe, December 21, 2017.

Air Force Instruction 16-1007, *Management of Air Force Operational Training Systems*, October 1, 2019.

Akkila, Alaa N., Abdelbaset Almasri, Adel Ahmed, Naser Al-Masri, Yousef Abu Sultan, Ahmed Y. Mahmoud, Ihab Zaquot, and Samy S. Abu-Naser, "Survey of Intelligent Tutoring Systems up to the End of 2017," *International Journal of Academic Information Systems Research*, Vol. 3, No. 4, April 2019.

Austin, Lloyd J., "Senate Armed Services Committee Advance Policy Questions for Lloyd J. Austin Nominee for Appointment to be Secretary of Defense," January 19, 2021. As of June 16, 2023:
https://www.armed-services.senate.gov/imo/media/doc/Austin_APQs_01-19-21.pdf

Betts, Richard K., *Military Readiness: Concepts, Choices, Consequences*, Brookings Institution, 1995.

Brown, Charles Q., Jr., and David H. Berger, "Redefine Readiness or Lose," War on the Rocks, March 15, 2021.

Crichton, Margaret Thomson, "From Cockpit to Operating Theatre to Drilling Rig Floor: Five Principles for Improving Safety Using Simulator-Based Exercises to Enhance Team Cognition," *Cognition Technology & Work*, Vol. 19, No. 1, 2017.

Defense Information Systems Agency and National Security Agency Zero Trust Engineering Team, *Department of Defense (DoD) Zero Trust Reference Architecture*, July 2022.

Department of Defense Instruction 8500.01, *Cybersecurity*, March 14, 2014, incorporating change 1, October 7, 2019.

Department of Defense Instruction 8510.01, *Risk Management Framework for DoD Systems*, July 19, 2022.

Dignitas Technologies, "Dignitas Awarded Flexible and Live Adaptive Training Tools (FLATT) Phase II SBIR," webpage, February 8, 2021. As of August 12, 2021: https://www.dignitastechnologies.com/flatt-ii-award

Folsom-Kovarik, J. T., Joel Sieh, and Anne M. Sinatra, "Reasoning About Team Roles and Responsibilities for Team Assessment," in Benjamin S. Goldberg, ed., *Proceedings of the Ninth Annual GIFT Users Symposium*, May 2021.

GIFT Tutoring, "Generalized Intelligence Framework for Tutoring (GIFT)," homepage, undated. As of August 12, 2021: https://www.gifttutoring.org/projects/gift/wiki/Overview

Harrison, Todd, "Rethinking Readiness," *Strategic Studies Quarterly*, Vol. 8, No. 3, Fall 2014.

Harrison, Steven, and Elyse Burmester, "FLATT: A New Real-Time Assessment Engine Powered by Gift," in Benjamin S. Goldberg, ed., *Proceedings of the Ninth Annual GIFT Users Symposium*, May 2021.

Herrera, G. James, *The Fundamentals of Military Readiness*, Congressional Research Service, R46559, 2020.

Jensen, Randy, Bart Presnell, Jeanine DeFalco, and Gregory Goodwin, "Designing a Distributed Trainer Using GIFT for Team Tutoring in Command Level Decision Making and Coordination," in Benjamin S. Goldberg, ed., *Proceedings of the Ninth Annual GIFT Users Symposium*, May 2021.

Johnston, Joan H., Anne M. Sinatra, and Zachari Swiecki, "Application of Team Training Principles to Visualizations for After-Action Reviews," in Anne M. Sinatra, Arthur C. Graesser, Xiangen Hu, Benjamin Goldberg, and Andrew J. Hampton, eds., *Design Recommendations for Intelligent Tutoring Systems*, Vol. 8: *Data Visualization*, U.S. Army Combat Capabilities Development Command, Soldier Center Simulation and Training Technology Center, 2020.

Joint Publication 1, *Doctrine for the Armed Forces of the United States*, Joint Chiefs of Staff, incorporating change 1, July 12, 2017.

Joint Publication 6-0, *Joint Communications System*, June 10, 2015, Joint Chiefs of Staff, incorporating change 1, October 4, 2019.

Koslowski, Steve W. J., and Georgia T. Chao, "Unpacking Team Process Dynamics and Emergent Phenomena: Challenges, Conceptual Advances, and Innovative Methods," *American Psychologist*, Vol. 74, No. 4, May–June 2018.

Kurian, John, "Simulators Division (AFLCMC/WNS) Update," briefing slides, June 16, 2021. As of August 19, 2021:
https://www.trainingsystems.org/-/media/sites/ntsa/events/2021/11t0/presentations/wed-16-june/16-june_0845_kurian_air-force-simulators-division-update.ashx

MacMillan, Jean, Eileen B. Entin, and Rebecca Morley, "Measuring Team Performance in Complex and Dynamic Military Environments: The SPOTLITE Method," *Military Psychology*, Vol. 25, No. 3, 2013.

Mane, Muharrem, Anthony D. Rosello, Paul Emslie, Thomas Edward Goode, Henry Hargrove, and Tucker Reese, *Developing Operationally Relevant Metrics for Measuring and Tracking Readiness in the U.S. Air Force*, RAND Corporation, RR-A315-1, 2020. As of August 12, 2021:
https://www.rand.org/pubs/research_reports/RRA315-1.html

Mansikka, Heikki, K. Virtanen, D. Harris, and M. Jalava, "Measurement of Team Performance in Air Combat—Have We Been Underperforming?" *Theoretical Issues in Ergonomics Science*, Vol. 22, No. 3, 2021.

Marlow, Shannon L., Christina N. Lacerenza, Jensine Paoletti, C. Shawn Burke, and Eduardo Salas, "Does Team Communication Represent a One-Size-Fits-All Approach? A Meta-Analysis of Team Communication and Performance," *Organizational Behavior and Human Decision Processes*, Vol. 144, 2018.

Martin, Bradley, Michael E. Linick, Laura Fraade-Blanar, Jacqueline Gardner Burns, Christy Foran, Krista Romita Grocholski, Katherine C. Hastings, Sydney Jean Litterer, Kristin F. Lynch, and Jared Mondschein, *Measuring Strategic Readiness: Identifying Metrics for Core Dimensions*, RAND Corporation, RR-A453-1, 2021. As of August 17, 2021:
https://www.rand.org/pubs/research_reports/RRA453-1.html

Mattis, Jim, *Summary of the 2018 National Defense Strategy of the United States of America: Sharpening the American Military's Competitive Edge*, U.S. Department of Defense, 2018. As of August 12, 2021:
https://dod.defense.gov/Portals/1/Documents/pubs/2018-National-Defense-Strategy-Summary.pdf

McDaniel, Susan H., and Eduardo Salas, "The Science of Teamwork: Introduction to the Special Issue," *American Psychologist*, Vol. 73, No. 4, 2018.

Moulton, Seth, Jim Banks, Susan Davis, Scott DesJarlais, Chrissy Houlahan, Paul Mitchell, Elissa Slotkin, and Michael Waltz, *Future of Defense Task Force Report 2020*, House Armed Services Committee, 2020.

Normand, Sean L., and Joan H. Johnston, "A Qualitative Study on Behavioral Markers of Team Cohesion and Collective Efficacy to Inform the Army's Synthetic Training Environment," paper presented at 2020 Virtual Interservice, Industry, Training, Simulation, and Education Conference (vIITSEC), National Training and Simulation Association, 2020.

Paul, Christopher, Jessica Yeats, Colin P. Clarke, Miriam Matthews, and Lauren Skrabala, *Assessing and Evaluating Department of Defense Efforts to Inform, Influence, and Persuade: Handbook for Practitioners*, RAND Corporation, RR-809/2-OSD, 2015. As of August 20, 2021:
https://www.rand.org/pubs/research_reports/RR809z2.html

Reising, Deanna L., Douglas E. Carr, Sally Gindling, Roxie Barnes, Derrick Garletts, and Zulfukar Ozdogan, "An Analysis of Interprofessional Communication and Teamwork Skill Acquisition in Simulation," *Journal of Interprofessional Education & Practice*, Vol. 8, No. 1, September 2017.

Rumbaugh, Russell, *Defining Readiness: Background and Issues for Congress*, Congressional Research Service, R44867, June 14, 2017.

"Simulator Concurrency: Why Military Operators Know It's Important to Winning the Fight," webpage, Modern Military Training, March 30, 2021. As of June 16, 2023:
http://modernmilitarytraining.com/training-effectiveness/simulator-concurrency-why-military-operators-know-its-important-to-winning-the-fight/

Smith, Ebb, Jonathan Borgvall, and Patrik Lif, *Team and Collective Performance Measurement*, DSTL, Policy and Capability Studies, June 30, 2007. As of August 12, 2021:
https://apps.dtic.mil/sti/pdfs/ADA474089.pdf

Sottilare, Robert A., C. Shawn Burke, Eduardo Salas, Anne M. Sinatra, Joan H. Johnston, and Stephen B. Gilbert, "Designing Adaptive Instruction for Teams: a Meta-Analysis," *International Journal of Artificial Intelligence in Education*, Vol. 28, June 2017.

Spain, Randall, Wookhee Min, Jason Saville, Keith Brawner, Bradford Mott, and James Lester, "Automated Assessment of Teamwork Competencies Using Evidence-Centered Design-Based Natural Language Processing Approach," in Benjamin S. Goldberg, ed., *Proceedings of the Ninth Annual GIFT Users Symposium*, May 2021.

Stevens, Ron, Ryan Mullins, Xiangen Hu, Diego Zapata-Rivera, and Trysha Galloway, "Visualizing the Momentary Neurodynamics of Team Uncertainty," in Anne M. Sinatra, Arthur C. Graesser, Xiangen Hu, Benjamin Goldberg, and Andrew J. Hampton, eds., *Design Recommendations for Intelligent Tutoring Systems*, Vol. 8: *Data Visualization*, U.S. Army Combat Capabilities Development Command, Soldier Center Simulation and Training Technology Center, 2020.

Swiecki, Zachari, Morten Misfeldt, Xiangen Hu, and David Williamson Shaffer, "Visualizing Team Processes Using Epistemic Network Analysis: Affordances for Researchers, Educators, and Teams," in Anne M. Sinatra, Arthur C. Graesser, Xiangen Hu, Benjamin Goldberg, and Andrew J. Hampton, eds., *Design Recommendations for Intelligent Tutoring Systems*, Vol. 8: *Data Visualization*, U.S. Army Combat Capabilities Development Command, Soldier Center Simulation and Training Technology Center, 2020.

Toukan, Mark, Matthew Walsh, Ajay K. Kochhar, Emmi Yonekura, and David Schulker, *Air Force Operational Test and Training Infrastructure: Barriers to Full Implementation*, RAND Corporation, RR-A992-1, 2022. As of June 15, 2023: https://www.rand.org/pubs/research_reports/RRA992-1.html

USAF—*See* U.S. Air Force.

U.S. Air Force, "*Air Force Operational Training Infrastructure 2035 Flight Plan*, September 5, 2017.

U.S. Department of Defense, *Department of Defense Modeling and Simulation (M&S) Glossary*, Modeling and Simulation Coordination Office, October 1, 2011.

U.S. Government Accountability Office, *Department of Defense Domain Readiness Varied from Fiscal Year 2017 Through Fiscal Year 2019*, GAO-21-279, April 2021. As of August 12, 2021: https://www.gao.gov/assets/gao-21-279.pdf

Vatral, Caleb, Naveeduddin Mohammed, Gautam Biswas, and Benjamin S. Goldberg, "Gift External Assessment Engine for Analyzing Individual and Team Performance for Dismounted Battle Drills," in Benjamin S. Goldberg, ed., *Proceedings of the Ninth Annual GIFT Users Symposium*, May 2021.

Vocavio Technologies, "Vsim: Pilot Training," webpage, undated. As of August 12, 2021: https://vocavio.com/our-solutions/vsim-pilot-training

Vonthoff, Tony, "The Importance of Fidelity in Simulation and Training," webpage, Modern Military Training, August 22, 2017. As of August 12, 2021: http://modernmilitarytraining.com/training-realism/importance-fidelity-simulation-training/

Wolfe, Frank, "U.S. Air Force Joint Simulation Environment for F-35A, F-22, Other Platforms to Build on F-35 JSE," *Defense Daily*, July 23, 2021.

Yates, Nick, "OTTI Update: Synthetic Test and Training Capability," NTSA Simulation & Training Community Forum 2021, August 11, 2021.